Einstein's Inertial Field

Franklin Felber

Einstein's Inertial Field

How it accounts for the Pioneer anomaly, MOND, Hubble redshift, cosmic 'acceleration' and Schrödinger equation

With 55 Illustrations

Published by Starmark Physics
Starmark, Inc.
P. O. Box 270710
San Diego, California 92198
USA

Felber, Franklin
    Einstein's inertial field : how it accounts for the Pio-
    neer anomaly, MOND, Hubble redshift, cosmic 'accel-
    eration' and Schrödinger equation

    / Franklin Felber.
    Includes bibliographical references.
    ISBN-13: 978-1530313662
    Keywords: Inertial field · Static universe · Pioneer
    anomaly · MOND · Hubble redshift · Cosmic accelera-
    tion

First Edition
    First Printing   April 2016

Printed in the United States of America.

ISBN-13: 978-1530313662
ISBN-10: 153031366X

Contents

Preface

In one sense at least, this book contains no new physics. Everything in this book is based more or less directly on a century-old equation, Einstein's equation of general relativity.

What you will find to be new about this book is a set of exact solutions of Einstein's equation. This new set of exact solutions reveals a wealth of information about our universe and seems to account for a wide variety of anomalous phenomena that have been observed over the past several decades. These phenomena range in scale from our solar system to the farthest reaches of the cosmos. The new solutions also strongly suggest physical foundations for such classical physics as Newton's third law of motion and even deterministic foundations for the quantum physics of Schrödinger's wave equation. A technical abstract for this book would read like this:

New exact solutions of Einstein's equation for a static universe with discrete masses reveal the existence of an inertial field. The new solutions feature an inertial time dilation that results in an event horizon in a static, nearly flat universe at a radius of 13.8 ± 1.2 Gly. Associated with the inertial field is a 4-vector inertial-drag force proportional to an invariant scalar acceleration. The inertial field accounts for anomalies at the scale of our solar system and at galactic and cosmological scales. Inertial time dilation accounts for the constant component of the Pioneer anomaly, but has no discernible effect on the precession rate of planetary orbits. In clusters and galaxies, the inertial field accounts indirectly for such effects as flat rotation curves, the Tully-Fisher relation, and the limit on mean surface brightness of spiral galaxies. The effect of the inertial drag force on hot plasmas offers an alternative interpretation of the Bullet cluster. The inertial field also accounts for the Hubble redshift and the appearance of cosmic acceleration in a static Einstein universe. With no adjustable parameters, inertial time dilation and inertial drag account completely for Type Ia supernova pulse data in a static universe. Acceleration of a mass produces dipole inertial radiation. Fluctuations of the cosmic inertial radiation background can account for all the quantum mechanical effects described by the Schrödinger wave equation. The power spectral density of inertial noise is calculated. The possibility that the inertial field is quantized and might be mediated by a massless spin-1 boson is discussed.

The book is called *Einstein's Inertial Field* because it is clear that Einstein must have understood that the bases of the physical laws governing inertia were already embodied in his theory of general relativity. The new solutions in this book show how the gravitational field of his theory is just the inertial field of the universe in close proximity to a mass. And the inertial field of the universe is just the sum of gravitational contributions of all the mass in the universe. These points are discussed in more detail in the Introduction and are supported mathematically by the exact solutions in Ch. 3.

Regardless of your background in physics, the recommended starting point for reading this book is Ch. 1, "Introduction," followed by Ch. 15, "Key Results." Both chapters are equation-free zones. Reading both will give a good understanding of the new perspective offered by these new solutions of Einstein's equation.

Next, you are invited to read App. F, which is the only other equation-free zone in the book. Appendix F may be more accurately characterized as philosophy of science rather than physics. The new solutions suggest that Einstein's inertial field can account for a wide variety of phenomena in a *static* universe, one that is neither expanding nor contracting. Appendix F addresses the obvious question of how a static universe can come to be and whether such a question can even be posed in a scientific context.

No matter how questions of the origin of the universe might ultimately be resolved, the exciting news now is that we are on the cusp of discovering just what kind of universe we live in. Direct measurements of cosmic acceleration are becoming feasible. Direct measurements will soon be able to determine which of two models, the Einstein static universe or the Big Bang accelerating universe, if either, is correct. Although the data will take the order of a decade to compile, the results should determine once and for all whether the universe is static, as Einstein first suggested, or expanding with a cosmic acceleration, as Nobel-prize-winning studies of supernovas now seem to suggest. Or some other possibility that may not even have been considered yet.

After the introduction and conclusions, the reader with some background in physics might enjoy scanning all the other equation-laden chapters, looking for blocks of text, particularly in introductory paragraphs. The text is intended to support and give physical insight into the mathematical development. And it is not necessary to follow the math to gain the insights.

Lastly, the diligent reader can tackle all the chapters between 1 and 15 in full detail. Even here, though, an effort was made to develop the equations step-by-step. No exercises are left to the reader. If the step-by-step development requires so many equations that they distract, rather than illuminate, then the equations and their step-by-step development are bundled into one of the five mathematical appendices for the truly assiduous reader.

Whether you scan this book casually or dive deeply into the math, the hope is that you will come to appreciate that Einstein's century-old masterpiece, his original theory of general relativity with no modifications, still has a world of new insights to offer into our understanding of our universe.

Chapter 1
Introduction

Two years before he finished formulating general relativity, Einstein was grati-
fied to find that certain aspects of inertia already seemed to be embodied in his
incipient theory. In particular, Einstein noted in his letter [1, 2] to Ernst Mach
that translational and rotational accelerations of a spherical shell of matter would
produce an accelerative force (*"beschleunigende Kraft"*) on a particle within the
shell. This remarkable departure from Newton's shell theorem seemed to Einstein to be consonant with Mach's idea that distant mass influences local inertial
frames and particle dynamics.

Surely, Einstein reasoned as well that if an acceleration of a shell produces a
gravitation-related accelerative force on a particle inside, then an acceleration of
a particle must produce an accelerative force on a shell outside, that is, on mass
within the future light cone of the particle. And by Newton's Third Law, "To
every action there is always opposed an equal reaction ...," there must be a
gravitation-related reaction force on the accelerating particle. Since the mass in
the future light cone is distant, but the reaction force on the accelerating particle
is instantaneous, then the gravitation-related reaction force must be mediated by
a field that is present everywhere.

This book finds that the underpinnings of inertia are in fact embedded within Einstein's equation of general relativity. Exact solutions of Einstein's equation for a static, homogeneous universe in Ch. 2 reveal the existence of an inertial field and an associated inertial force in Ch. 6. Extending the exact solutions
of Ch. 2 to a static, 'homogeneous' universe of discrete masses in Ch. 3 further
reveals that the gravitational field of a mass is just the local contribution of that
mass to the inertial field of the universe. And near a mass, the inertial field
takes on the appearance of a gravitational field. What Wheeler called Sciama's
'sum for inertia' [3] is found to be closely related to an exact solution of Einstein's equation for a universe of discrete masses.

Just as Einstein's early intuition regarding the intrinsic relation of inertia to
gravitation deserves a closer look, so does his early intuition regarding cosmology. Einstein's original cosmological model, now known as the Einstein static
universe, has been largely, though not completely, abandoned by the cosmology

community. With reluctance, Einstein himself abandoned the model of a static universe in light of the discovery by Hubble of a nearly linear relationship with distance of the redshift of light from distant astronomical objects. Today this Hubble redshift is almost universally attributed to an expansion of our universe and is thought to be absent in a static universe.

Instead, Ch. 4 shows that the Hubble redshift can be attributed to a dilation of time that is a feature of a *static* universe. Moreover, the combined effects of this inertial time dilation with the dissipative effects of the inertial force are found in Ch. 5 to account for the appearance of cosmic acceleration in a *static* universe.

The inertial field accounts not only for these observations on the cosmological scale, but for observations of anomalies that have been made at the scale of our solar system and at galactic scales, as well. As indicated in Table 1.1, the inertial field characterized by this book is consistent with, and may even account for, the inertial forces of Newton's third law at the laboratory scale and all other scales, as shown in Ch. 12, and may account for all the quantum mechanical effects described by the Schrödinger wave equation at atomic and sub-atomic scales, as shown in Ch. 13. The modified Newtonian dynamics (MOND) and lambda-cold-dark-matter (ΛCDM) models, on the other hand, even with their adjustable parameters, can account for only a subset of phenomena at cosmological and galactic length scales.

Chapter 11 shows that an acceleration of masses in gravitationally unbound quadrupoles produces dipole inertial radiation in the far (radiation) zone. Chapter 14 discusses whether this dipole inertial radiation may be transported to the far zone by a massless spin-1 boson, just as electromagnetic radiation is.

Chapter 2, "Exact Inertial Field of Static, Homogeneous Universe," presents a family of exact solutions of Einstein's equation for a static, homogeneous spacetime. One member of this family, the Friedmann-Lemaître model of a static universe, is a trivial solution for which the time-dilation factor of the spacetime metric has been suppressed by the choice of coordinates. The time-dilation factor gives the apparent slowing of clocks as a function of range and scalar curvature of a static universe. Whether our universe is expanding or not, this inertial time dilation causes clocks to appear to stop completely at some finite range R. Appendix A presents the detailed derivation of the exact solutions presented in Ch. 2.

Table 1.1. An inertial field in a nearly flat, static universe accounts for physical phenomena on the scale of our universe, galaxies, our solar system, and on laboratory and atomic scales.

SCALE OF PHENOMENA

MODEL	Universe	Galaxy	Solar System	Lab	Atomic
ΛCDM	✓	✓			
MOND		✓			
Inertial Field	✓	✓	✓	✓	✓

Chapter 3, "Exact Inertial Field of Static Universe with Discrete Masses," generalizes the exact solutions in Ch. 2 of Einstein's equation for a static, homogeneous spacetime to include a spherically-symmetric central mass, and then to include a discrete set of masses rather than a continuous mass density. The generalized solutions suggest, conceptually at least, that the gravitational field of a mass is just the contribution of the mass to the inertial field of the universe.

Chapter 4, "Hubble Redshift and Inertial Drag in a Static Universe," calculates the exact frequency redshift and dissipative work done by light as it propagates through the inertial field of a static universe. Just as a mass delivers a transverse gravitational impulse to light, light delivers a transverse gravitational impulse, mediated by the inertial field, to all the mass in its future light cone. The redshift parameter z for a static, homogeneous universe is calculated exactly as a function of the inertial time dilation derived in Ch. 2.

Chapter 5, "Appearance of Cosmic 'Acceleration' in a Static Universe," calculates the departure from a linear cosmological redshift that occurs as a consequence of an inertial field acting on light propagating through a *static*, homogeneous universe. In a static universe, the effects on the redshift parameter of an inertial field can appear to have been caused by an acceleration of an expanding universe. The redshift and energy and power loss caused by an inertial field in a nearly flat *static universe* is shown here to account fully, with no adjustable parameters, for observations of the luminosity and distance modulus of Type Ia supernovas (SN-Ia).

Chapter 6, " Exact Particle Motion in a Static Inertial Field," derives the covariant equation of motion and calculates the exact motion of a particle in the inertial field of a static, homogeneous universe. The inertial field exerts a drag force on any particle moving with respect to the static field. This dissipative drag force is proportional to a universal inertial drag constant, $D = 0.69 \pm 0.06$ nm / s^2, an invariant scalar equal to Hubble's constant times the speed of light.

Chapter 7, "Inertial Time Dilation and the Pioneer Anomaly," shows how inertial time dilation in a static, homogeneous universe can account for the Pioneer anomaly, an apparent acceleration of the Pioneer 10 and Pioneer 11 spacecraft directed towards the observer. Inertial time dilation, with no adjustable parameters, accounts for the constant component of the apparent acceleration, but not for the smaller, slowly decaying component, which may be caused by effects of spacecraft thermal design.

Chapter 8, "Effects of Weak Inertial Drag on Orbital Motion," finds that, although inertial drag does cause the orbital radius, angular momentum, and energy of planetary orbits to decay, inertial drag does not cause any discernible precession of the periapsis of the orbits. Thus, the results are consistent with precise measurements of precession of planetary orbits in our solar system. Appendix B presents the detailed derivation of the calculations of Ch. 8.

Chapter 9, "Flat Rotation Curves and Tully-Fisher Relation in Clusters, Galaxies, and Gases," shows that the inertial field can account indirectly for such effects of modified Newtonian dynamics (MOND) as the Tully-Fisher relation and flat rotation curves in clusters and galaxies, and the limit on mean surface brightness of spiral galaxies, known as Freeman's law. The inertial field accounts for MOND effects only indirectly because inertial drag causes the outer regions of spiral galaxies to be pressure-supported, rather than rotation-supported, and MOND effects follow in pressure-supported systems. Appendix C calculates geometrical effects on the gravitational field of a disk galaxy *vs.* a spherical galaxy.

Chapter 10, "Gravitational Reach and Inertial Drag in Cluster, Galaxy, and Gas Dynamics," calculates effects of inertial drag on gravitational encounters of particles with much larger masses, such as the encounters of gas clouds with the core of a galaxy. This section also models the deceleration by inertial drag of pressure-supported systems, a calculation that offers an alternative interpretation of the 'bow-shock' shape of the Bullet cluster plasma cloud.

Chapter 11, "Dipole Inertial Radiation from Unbound Quadrupoles," shows that dipole gravitational disturbances from gravitationally *unbound* mass quadrupoles propagate to the radiation zone with signal strength at least of quadrupole order. Despite having constant mass dipole moment, unbound quadrupoles nevertheless produce dipole perturbations of inertial fields that do not completely destructively interfere as they propagate to the radiation zone. Angular distributions of parallel-polarized and transverse-polarized dipole inertial radiation are calculated for a harmonic oscillator affixed to a much heavier mass. Appendix D presents the derivation from [4] of the weak gravitational field of a particle undergoing arbitrary relativistic motion, exact to all orders of source velocity.

Chapter 12, "Dipole Inertial Radiation, Inertial Forces, and Newton's Third Law," calculates the dipole inertial radiation produced by a solitary unbound quadrupole during accelerated motion. The calculation shows that the dipole inertial field produced by an accelerating mass in the radiation zone is consistent with an inertial force acting instantaneously to oppose the acceleration, and is thereby consistent with the reaction force of Newton's third law.

Chapter 13, "Dipole Inertial Noise and the Schrödinger Wave Equation," shows that dipole inertial noise can account classically for all the quantum mechanical effects described by the time-dependent Schrödinger wave equation, if the power spectral density of the noise is about $l_p^2\omega^3$, where l_p is the Planck length and ω is the angular frequency of the noise, and if the inertial noise spectrum is cut off at about the Planck frequency. Appendix E calculates general statistical properties of a noise field, applicable to the inertial noise field.

Chapter 14, "Is the Inertial Field Mediated by a Massless Spin-1 Boson?," discusses the possibility that the inertial field, which exerts a vector force on particles as the electromagnetic field does, might be mediated by a massless spin-1 boson like the photon.

Chapter 15, "Key Results," summarizes the key results of the book. Appendix F addresses the question of which matters are appropriate subjects of scientific inquiry according to the falsifiability criterion of [5], particularly with respect to the origin of a static universe.

Chapter 2
Exact Inertial Field of Static, Homogeneous Universe

This section summarizes an exact solution in isotropic Cartesian coordinates of Einstein's equation for a static (time-independent and time-reversible), homogeneous spacetime. Appendix A presents the detailed calculations of the new solution. The Friedmann-Lemaître model of a static universe is shown to be a trivial solution of the static field equations. An exact solution for a static, homogeneous spacetime including discrete masses is presented in Ch. 3.

In isotropic Cartesian coordinates, x, y, z, the most general static, homogeneous spacetime interval is

$$c^2 d\tau^2 = e^{P(\mathbf{x})} c^2 dt^2 - e^{Q(\mathbf{x})} (dx^2 + dy^2 + dz^2), \qquad (2.1)$$

where P and Q are functions of $\mathbf{x} - \mathbf{x_0}$, the displacement 3-vector from any origin, $\mathbf{x_0}$. Einstein's equation is

$$R^{\mu}_{\ \nu} = \left(C/2 - \Lambda/c^2 \right) \delta^{\mu}_{\ \nu} - \kappa T^{\mu}_{\ \nu}, \qquad (2.2)$$

where $R^{\mu}_{\ \nu}$ is the Ricci tensor, $C \equiv R^{\sigma}_{\ \sigma}$ is the curvature scalar, Λ is the cosmological constant, $\delta^{\mu}_{\ \nu}$ is the four-dimensional Kronecker delta, $T^{\mu}_{\ \nu}$ is the energy-momentum tensor, and $\kappa \equiv 8\pi G/c^4$.

As calculated in Appendix A, Einstein's equation in isotropic coordinates in a static spacetime is exactly given by two coupled second-order differential equations in Q and P,

$$4\nabla^2 Q + (\nabla Q)^2 = -kCe^{Q} \qquad (2.3a)$$

$$2\nabla^2 P + (\nabla P)^2 + (\nabla P \cdot \nabla Q) = -(2-k)Ce^{Q}, \qquad (2.3b)$$

where the scale factor k is defined as $k \equiv 2(1 - 2R^0_{\ 0}/C)$. The $T^0_{\ 0}$ component of the energy-momentum tensor is given by the 'scaled curvature', $kC/4 = \kappa T^0_{\ 0} + \Lambda/c^2$. For a static, homogeneous spacetime, the solution of Eq.

(2.3a) for which $e^Q = 1$ at $\mathbf{x} = \mathbf{x}_0$ is

$$e^Q = (1 + kCr^2 / 48)^{-2},$$ (2.4)

where $\mathbf{r} = \mathbf{x} - \mathbf{x}_0$ is the displacement 3-vector from the origin. Different values of k correspond to different coordinate transformations of the metric. In Appendix A, three distinct coordinate transformations are considered, corresponding to: $R^0_{\ 0} = 0$ and $k = 2$; $R^0_{\ 0} = C / 4$ and $k = 1$; and $R^0_{\ 0} = C / 2$ and $k = 0$.

If $R^0_{\ 0} = 0$, then $k = 2$, and a trivial solution of Eq. (2.3b) is $e^P = 1$, which leads to the Friedmann-Lemaître model of a homogeneous universe with the cosmological constant Λ adjusted to give a static solution. The general solution in spherical coordinates of Eq. (2.3b), calculated in Appendix A, is

$$e^{P/2} = \alpha_1 + (\alpha_2 / r)(1 - r^2 / R^2),$$ (2.5)

where α_1 and α_2 are constants, and $R^2 \equiv 24 / C$.

If $R^0_{\ 0} = C / 4$, then $k = 1$, corresponding to an empty-Lemaître model of a static universe with $T^0_{\ 0} = 0$ and $\Lambda \neq 0$. For $R^2 \equiv 48 / C$, the general solution of Eq. (2.3b) is

$$e^{P/2} = \left(R^2 + r^2\right)^{-1} \left[\alpha_1 \left(R^2 - r^2\right) + (\alpha_2 / r)\left(1 - 6r^2 / R^2 + r^4 / R^4\right)\right].$$ (2.6)

If $R^0_{\ 0} = C / 2$, then $k = 0$, and for $C \geq 0$ and $R^2 \equiv 4\pi^2 / C$, the general solution of Eq. (2.3b) is

$$e^{P/2} = (\alpha_1 R / \pi r)\sin(\pi r / R) + (\alpha_2 / r)\cos(\pi r / R).$$ (2.7)

For $\alpha_1 = 1$ and $\alpha_2 = 0$ in Eqs. (2.5) – (2.7), the exact solutions in spherical coordinates for these three cases of a static, homogeneous spacetime with nonnegative curvature are shown in Fig. A.1 of Appendix A.

If $R^0_{\ 0} = C / 2$ and $k = 0$, but $C < 0$, corresponding to positive energy density, $T^0_{\ 0} > 0$, then the general solution of Eq. (2.3b) is

$$e^{P/2} = (\alpha_1 L / r)\sinh(r / L) + (\alpha_2 L / r)\cosh(r / L),$$ (2.8)

where $L \equiv (-2 / C)^{1/2}$ is a constant length. For $\alpha_1 = 1$ and $\alpha_2 = 0$, a solution for a static, homogeneous spacetime with $R^0_{\ 0} = C / 2 < 0$ is

$$e^{P} = (L/r)^2 \sinh^2(r/L).\tag{2.9}$$

The apparent singularities at $r = 0$ (for $\alpha_2 \neq 0$) in Eqs. (2.5) – (2.8) are an artifact of imposing spherical coordinates on a homogeneous universe that has no natural origin. A general solution of Eq. (2.3b) without singularities can be found in Cartesian coordinates. The solution is valid along every ray from the origin. The g_{00} component of the metric of a static, homogeneous spacetime for an observer at the origin can then be defined through the solution of Eq. (2.3b) along the set of all rays from the origin. In the following then, x represents a measure in a Cartesian coordinate of the distance from the origin along any ray, as represented in Fig. 2.1. And the Laplacian operator along any ray is d^2/dx^2.

In terms of this Cartesian coordinate x, the spacetime interval is

$$c^2 d\tau^2 = e^{P(x)} c^2 dt^2 - e^{Q(x)} (dx^2 + dy^2 + dz^2),\tag{2.10}$$

and Einstein's equations, Eqs. (2.3), become

$$d^2 q / dx^2 = -kCq^5 / 16\tag{2.11a}$$

$$d^2 w / dx^2 = -(8 - 3k)Cq^4 w / 16,\tag{2.11b}$$

where $q \equiv e^{Q/4}$ and $w \equiv qe^{P/2}$. The Emden-Fowler equation, Eq. (2.11a), has an exact solution in Bessel-functions of fractional order [7]. But for $k = 0$, corresponding to $R^0_{\ 0} = C/2$ and a flat 3-volume, a simple and preferred solution of Eq. (2.11a) is $e^{Q} = 1$. With this choice of coordinates, hereinafter called '*natural*' Cartesian coordinates, Eq. (2.11b) has the general solution along any ray from the origin,

$$e^{P/2} = \alpha_1 \sinh(C_0 x / R) + \alpha_2 \cosh(C_0 x / R),\tag{2.12}$$

where $C_0 \equiv (|C| R^2 / 2)^{1/2}$. Then the exact diagonal metric of the inertial field of a static, homogeneous universe in natural Cartesian coordinates may be expressed along any ray from the origin as

$$g_{00} = \frac{\sinh^2[C_0(1 - x / R)]}{\sinh^2 C_0}, \quad g_{11} = g_{22} = g_{33} = -1.\tag{2.13}$$

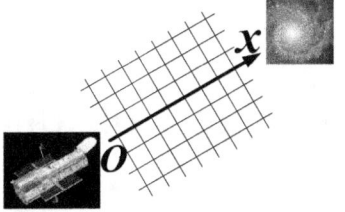

Fig. 2.1. Coordinate x of inertial-field metric measures Cartesian range along any ray from observer at origin. Photos from [6].

For $C > 0$, the hyperbolic sine functions in Eq. (2.13) are replaced by sine functions of the same arguments.

Figure 2.2 shows the exact g_{00} component of the diagonal metric of a static, homogeneous spacetime for several values of curvature and for a flat 3-volume, $g_{11} = g_{22} = g_{33} = -1$. This solution is one-dimensional along a ray, and the x coordinate is Cartesian and not curvilinear, as illustrated in Fig. 2.1. As shown in Fig. 2.2, for $C_0 \ll 1$, the spacetime interval,

$$c^2 d\tau^2 \approx (1 - x/R)^2 c^2 dt^2 - (dx^2 + dy^2 + dz^2), \tag{2.14}$$

is nearly independent of curvature, whether the curvature is positive or negative.

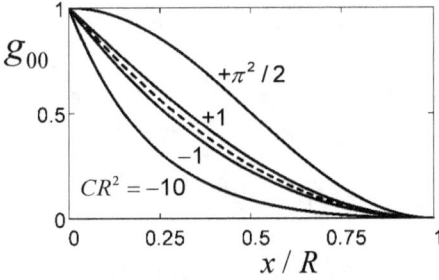

Fig. 2.2. Exact inertial field of static, homogeneous universe. Metric component g_{00} vs. normalized Cartesian coordinate x/R for normalized curvatures CR^2 indicated and for $g_{11} = g_{22} = g_{33} = -1$. Dashed curve is $C = 0$.

Chapter 3
Exact Inertial Field of Static Universe with Discrete Masses

This section generalizes the exact solutions in Ch. 2 of Einstein's equation for a static, homogeneous spacetime to include a spherically-symmetric central mass, and then to include a discrete set of masses rather than a continuous mass density. The generalized solutions suggest, conceptually at least, that the gravitational field of a mass is just the contribution of the mass to the inertial field of the universe.

According to Birkhoff's theorem, the field of a central, spherically symmetric mass is the same outside the mass as if all the mass were concentrated at the center. The contribution of a static central mass to the T^0_0 component of the energy-momentum tensor of a static, homogeneous spacetime with a central mass, therefore, is $m_0 c^2 \delta^3(\mathbf{r})$, where $\delta^3(\mathbf{r})$ is the three-dimensional Dirac delta function. Then Einstein's equation, Eqs. (2.3), for a static, homogeneous spacetime with a central mass becomes

$$e^{-Q}[\nabla^2 Q + (\nabla Q)^2 / 4] = -\kappa m_0 c^2 \delta^3(\mathbf{r}) - R^0_0 + C/2 \quad , \tag{3.1a}$$

$$e^{-Q}[\nabla^2 P + (\nabla P)^2 / 2 + (\nabla P \cdot \nabla Q)/2] = \kappa m_0 c^2 \delta^3(\mathbf{r}) + 2R^0_0 - 2C \quad . \tag{3.1b}$$

In terms of the new variables, $p(\mathbf{x}) \equiv e^{P(\mathbf{x})/2}$ and $q(\mathbf{x}) \equiv e^{Q(\mathbf{x})/4}$, Eqs. (3.1) become

$$4(\nabla^2 q)/q^5 = -\kappa m_0 c^2 \delta^3(\mathbf{r}) - R^0_0 + C/2, \tag{3.2a}$$

$$2(q\nabla^2 p + 2\nabla p \cdot \nabla q)/(pq^5) = \kappa m_0 c^2 \delta^3(\mathbf{r}) + 2R^0_0 - 2C \quad . \tag{3.2b}$$

In the flat-3-volume coordinates for which $R^0_0 = C/2$, combining Eqs. (3.2) gives the two Poisson's equations,

$$\nabla^2 q = -4\pi r_S q^5 \delta^3(\mathbf{r}), \tag{3.3a}$$

$$\nabla^2(pq) = +4\pi r_S pq^5 \delta^3(\mathbf{r}) - Cpq/2, \tag{3.3b}$$

where $r_S \equiv Gm_0 / 2c^2$ is the Schwarzschild radius in isotropic coordinates. The solutions of Eqs. (3.3) give the exact diagonal spacetime metric in isotropic coordinates for the inertial field of a static, homogeneous universe with a central mass as

$$g_{00} = \left(1 + \frac{r_S}{r}\right)^{-2} \left(\frac{\sinh[C_0(1 - x/R)]}{\sinh C_0} - \frac{r_S}{r}\right)^2$$

$$g_{11} = g_{22} = g_{33} = -\left(1 + r_S / r\right)^4 \; , \tag{3.4}$$

where $C_0 \equiv (|C| R^2 / 2)^{1/2}$ is the curvature parameter of Ch. 2, and the Cartesian coordinate x is distinguished from the radial spherical coordinate r as a measure of distance along any ray from the origin, as in Eq. (2.13). The significance of the distinction between x and r will become apparent in following sections.

Figure 3.1 shows this exact solution of Einstein's equation outside a central mass in the inertial field of a static, homogeneous universe. This solution, which is independent of curvature for $C_0 = 0$, reduces to the Schwarzschild solution in isotropic coordinates in the limit $R \to \infty$, and to the inertial field solution of Eq. (2.14) in the limit $r_S = 0$ of no central mass. To first order in r_S / R, event horizons are located at radii of r_S and $R - r_S$.

To generalize this solution to a universe comprising a discrete set of masses, rather than a continuous mass density, the $T^0{}_0$ component of the energy-momentum tensor is taken to be

$$T^0{}_0(\mathbf{r}) = \sum_{i=0}^{\infty} m_i(\mathbf{r}_i) c^2 \delta^3 (\mathbf{r} - \mathbf{r}_i), \tag{3.5}$$

where $\mathbf{r}_i$ is the position vector to the ith mass, m_i. The mean energy density, $\bar{T}$, of this distribution of discrete masses is taken to be a constant over any volume large enough that the universe appears to be homogeneous on the scale of that volume. And we choose natural Cartesian (flat-3-volume) coordinates, as above, in which $\kappa \bar{T} = -\Lambda / c^2 = -C/3$. In these coordinates, Einstein's equation for a static, homogeneous spacetime with a distribution of discrete masses becomes

$$e^{-Q}[\nabla^2 Q + (\nabla Q)^2 / 4] = -\kappa T^0{}_0(\mathbf{r}), \tag{3.6a}$$

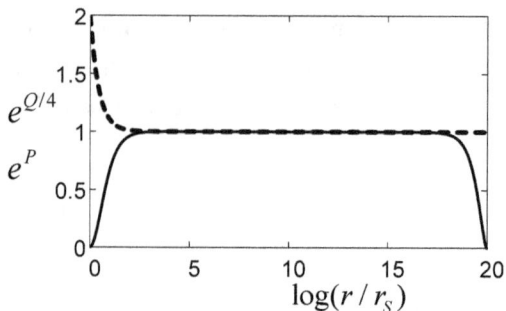

Fig. 3.1. Exact solution of Einstein's equation for inertial field of static, homogeneous universe with central mass, $e^{Q/4}$ (dashed) and e^P (solid) *vs.* log of radial distance (normalized to r_S) for $C_0 = 0$ and $R/r_S = 10^{20}$.

$$e^{-Q}[\nabla^2 P + (\nabla P)^2/2 + (\nabla P \cdot \nabla Q)/2] = 3\kappa\overline{T} + \kappa T^0_{\ 0}(\mathbf{r}) \quad . \tag{3.6b}$$

In terms of $p(\mathbf{x})$ and $q(\mathbf{x})$, Eqs. (3.6) become the Poisson's equations,

$$\nabla^2 q = -4\pi \sum_{i=0}^{\infty} Gm_i(\mathbf{r}_i) q^5 \delta^3(\mathbf{r} - \mathbf{r}_i)/2c^2 , \tag{3.7a}$$

$$\nabla^2 (pq) = (3\kappa\overline{T}/2)pq + 4\pi \sum_{i=0}^{\infty} Gm_i(\mathbf{r}_i) pq^5 \delta^3(\mathbf{r} - \mathbf{r}_i)/2c^2 \quad . \tag{3.7b}$$

If each mass is beyond the gravitational reach of its neighbors, conceptually the solution of Eq. (3.7a) is

$$q = 1 - \overline{S} + S(\mathbf{r}), \tag{3.8}$$

where

$$S(\mathbf{r}) \equiv \sum_{i=0}^{\infty} \frac{Gm_i(\mathbf{r}_i)}{2c^2 |\mathbf{r} - \mathbf{r}_i|} = \frac{Gm_0(\mathbf{r}_0)}{2c^2 |\mathbf{r} - \mathbf{r}_0|} + \frac{Gm_1(\mathbf{r}_1)}{2c^2 |\mathbf{r} - \mathbf{r}_1|} + \dots \tag{3.9}$$

is the 'sum for inertia', and $\overline{S}$ is the mean 'sum for inertia', which, like $\overline{T}$, is taken to be a constant over any volume large enough that the universe appears to be homogeneous on the scale of that volume.

If m_0 is taken to be a central mass located at $\mathbf{r}_0 = 0$, then in the neighborhood of the origin, the 'sum for inertia' becomes

$$S(\mathbf{r}) = \frac{r_S}{r} + \frac{Gm_1(\mathbf{r}_1)}{2c^2 |\mathbf{r} - \mathbf{r}_1|} + ... \approx \frac{r_S}{r} + \overline{S} \quad , \tag{3.10}$$

and the solution of Eq. (3.7a) becomes $q \approx 1 + r_S / r$. That is, the leading term in this expansion of the 'sum for inertia' is the term responsible for the Schwarzschild gravitational field, suggesting that the gravitational field of a mass is just the contribution of the mass to the inertial field of the universe, and is distinguishable from the inertial field only by the magnitude and proximity of the mass.

This derivation of the 'sum for inertia' suggests that the inertial field at a spacetime point is a sum of contributions from all the mass in the universe, or at least all the mass contained within the past light cone of the spacetime point. Viewed differently, one might say that the inertial field is what endows matter with its inertial properties, such as resisting acceleration with a force proportional to its mass. The ambiguity between mass underlying the inertial field and the inertial field endowing matter with mass is perhaps what Einstein was expressing by "...the division into matter and field is ... something artificial and not clearly defined [8]." This ambiguity led Einstein to the concept, or the hope, that, "We could regard matter as the regions in space where the field is extremely strong. ... A thrown stone is, from this point of view, a changing field, where the states of greatest field intensity travel through space with the velocity of the stone [8]." Einstein's view, "suggested by the great achievements of field physics," was, "There would be no place, in our new physics, for both field and matter, field being the only reality [8]."

Chapter 4
Hubble Redshift and Inertial Drag in a Static Universe

As light propagates through a static universe, it loses energy both by frequency redshift through inertial time dilation and by dissipative attenuation through inertial drag. Dissipative inertial drag does not affect the redshift parameter. The redshift parameter Z is calculated exactly as a function of the inertial time dilation derived in Ch. 2 for a static, homogeneous universe.

As light propagates across the universe, the inertial field affects its power and energy in three ways:

(i) Inertial time dilation causes light arriving at our sensors to be redshifted. In a universe such as ours, which has little curvature of spacetime, the fractional redshift from inertial time dilation is very nearly directly proportional to the distance travelled. In a static universe, inertial time dilation can account for the linear Hubble redshift that is attributed to a velocity of expansion proportional to distance.

(ii) Pulse stretching, like the time dilation redshift, is also caused by the apparent slowing of clocks at a distance. A light pulse emitted by a distant source appears to be dilated. Pulse stretching reduces the power of a light pulse arriving at our sensors, but not the pulse energy.

(iii) Inertial drag causes a loss of both energy and power from light propagating across the universe. The energy and power lost from light provides momentum and energy to mass lying in the future of the light to allow that mass to respond to the changing gravitational field of the light. In a universe like ours with little curvature, inertial drag causes the energy and power of a light pulse to decay exponentially with a half-life of about 9.6 ± 0.8 billion years.

The inertial field exerts a drag force on light propagating through the static field. Symmetry requires the drag force to be collinear with the Poynting vector, which points in the direction of propagation. Conservation of energy requires the drag force to oppose the Poynting vector.

The model of the universe for calculating inertial drag is of an infinite, static, homogeneous medium having no physical properties other than some constant linear combination of cosmological constant and energy density, which might include pressure.

As a pulse of light transports its momentum and energy through the universe, it modifies its gravitational potential with the rest of the mass of the universe. To do that, it must radiate enough energy to modify the potential and to cause the mass of the universe to respond to the new potential accordingly. The radiated energy must be of a magnitude and of a form that will cause isolated mass to move in response to the changed potential, even in the far (radiation) zone. That is, it must be dipole radiation. A pulse of light moving through a static universe is an example of an unbound quadrupole, and the dipole radiation of unbound quadrupoles, derived in [9], is discussed in Ch. 11 of this book.

For example, consider the deflection of a ray of starlight passing by the limb of the Sun. The starlight, having a longitudinal momentum p_x, acquires a transverse momentum $\Delta p_\perp \approx (4GM_\odot / bc^2) p_x$ after passing by the Sun, where $M_\odot$ is the mass of the Sun, and b is the impact parameter (distance of closest approach) of the ray of light with the center of the Sun. Easily overlooked is that by conservation of momentum the starlight delivers a transverse momentum impulse to the Sun of $-\Delta p_\perp$. For unbound quadrupoles like the starlight passing the Sun, the potential of the Sun's gravitational field is not conservative. The gravitational field of the Sun converts some of the longitudinal momentum and energy of passing starlight to transverse momentum and kinetic energy of the Sun.

When light propagates through the universe, some of its momentum and energy is similarly converted to momentum and kinetic energy of the rest of the mass of the universe. The following simple dimensional analysis, illustrated in Fig. 4.1, shows that when light with momentum p_x passes through a thin slice of the universe with thickness Δx, a fraction of its momentum and energy $\Delta p_\perp / p_x \sim \Delta x / R$ is converted to momentum and kinetic energy of mass in the universe.

Consider a toy model of the universe as a cylindrical shell of radius R and linear mass density μ. From the 'sum for inertia', $\mu \sim c^2 / G$. A slice of this cylindrical shell is a ring of thickness Δx and mass $\mu \Delta x$. Imparting momentum and energy to the ring, a pulse of light propagating a distance Δx along the axis of the cylinder loses fractional momentum and energy $\Delta p_\perp / p_x \sim G(\mu \Delta x) / Rc^2 \sim \Delta x / R$.

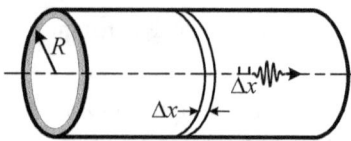

Fig. 4.1. Simple model of static universe shows fractional energy dissipation per distance Δx of propagation of a pulse of light is $(\Delta p / p) / \Delta x \sim 1 / R$.

The exact energy dissipation by inertial drag is calculated in Chs. 5 and 6. The exact redshift by inertial time dilation of light propagating through a static universe is derived as follows.

Because there is no preferred coordinate origin, Cartesian coordinates are most natural for calculating the redshift of light in a static, homogeneous universe. Consider an observer at rest at any coordinate origin looking in the $+x$ direction at a set of clocks and light sources arrayed along the x axis. If the clocks are at rest, then from the static, homogeneous spacetime interval of Eq. (2.1), the observer sees the clocks running slowly, that is, he sees their time intervals dilated by a factor $\dot{\imath} = e^{-P/2}$, where $\dot{\imath} \equiv dt / d\tau$. With the inertial-field metric given by Eq. (2.13), the time dilation of the clocks apparent to the observer depends upon their distance x from the origin as

$$\dot{\imath} = (\sinh C_0) / \sinh[C_0(1 - x / R)], \tag{4.1}$$

where $C_0 \equiv (|C| R^2 / 2)^{1/2}$ is the dimensionless curvature parameter from Eq. (2.13).

In Fig. 4.2, suppose a radio-frequency (rf) antenna at A produces a plane-wave pulse of radiation having a central frequency f_0, a pulse duration T_0, and time-integrated energy flux, also called energy fluence, E_0. The pulse either is reflected back to A from a reflector a short distance Δx away, as in Fig. 4.2(a), or is measured by B, reproduced identically, and then re-radiated back to A, as in Fig. 4.2(b). The two cases are equivalent.

Further suppose that the rf generators and their currents at A and B are optically visible to their observer counterparts, B and A respectively. For example, suppose that B can see the alternating current (AC) driving the antenna that produces the radiation pulse at A, and that each cycle of the AC corresponds to a wave cycle of the radiation pulse, so that the frequency of the AC and of the

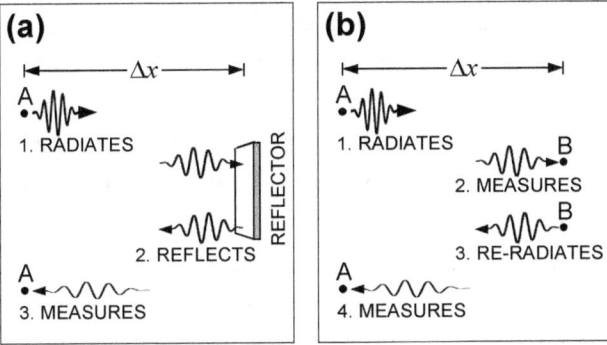

Fig. 4.2. *Gedankenexperiment* to derive the loss of energy and power of radiation between two observers at rest in a static inertial field.

antenna at A is f_0, the same as the central frequency of the pulse. Since both the optical image of the generator at A and the pulse produced by A travel at the same speed, each wave cycle of the pulse arrives at B simultaneously with the image of the AC cycle that produced it.

In Fig. 4.2, let the central frequency, pulse duration, and energy fluence of the pulse received at the reflector at B be denoted f_1, T_1, and E_1. And let the central frequency, pulse duration, and energy fluence of the pulse at its return to A be denoted f_2, T_2, and E_2.

By time dilation in an inertial field, B sees the AC generator at A slowed, just as a clock appears slowed. By causality, since each wave cycle of the pulse arrives simultaneously with the image of the AC cycle that produced it, the central frequency of the pulse appearing at B must be redshifted to $f_1 = f_0 / i(\Delta x)$. Similarly, the pulse duration measured at B must appear stretched to $T_1 = T_0 i(\Delta x)$.

The power radiated by the rf antenna at A is $R_{rad} I_0^2$, where R_{rad} is the radiation resistance of the antenna and I_0 is the root-mean-square (rms) current driving it. Since the current I_0 at A appears to B to be reduced to $I_1 = I_0 / i(\Delta x)$, the radiated power appears to B to be reduced by inertial time dilation and pulse stretching by a factor $[i(\Delta x)]^{-2}$, and the radiated energy, which is the time-integrated radiated power, appears to be reduced by inertial time dilation and pulse stretching by a factor $[i(\Delta x)]^{-1}$.

Because power scales inversely with time as t^{-3}, however, the power of a light pulse from a source at rest in a static universe measured by an observer at rest a short distance Δx away is reduced from the radiated power by a factor $[i(\Delta x)]^{-3}$, not $[i(\Delta x)]^{-2}$. The extra factor of $i(\Delta x)$ in the reduction of power is caused by inertial drag, as discussed in Ch. 5.

In the static gravitational field of a mass, such as a Schwarzschild field, the redshift of light climbing out of the gravitational potential well of the mass represents a reversible conversion of electromagnetic energy to potential energy. That is, if the light is reflected back to its source, it will regain that potential energy and be blueshifted back to its original frequency. In an inertial field, on the other hand, there is no localized potential well. When the pulse from A arrives redshifted at B, the loss of energy by frequency redshift cannot be regained.

The ratio of light frequencies at B and A is $f_1 / f_0 = [i(\Delta x)]^{-1}$. And when the pulse from A is returned either by a reflector or by B to its source at A, rather than regaining the expended energy, as in a gravitational field, the pulse loses further energy by time dilation. By symmetry in an inertial field, moreover, A and B can reverse the order of their operations with no difference between the two cases evident in the pulse returned to its source. By symmetry, therefore, the change of central frequency, pulse duration, and energy fluence on each leg of the round trip in an inertial field are related by $f_1 / f_0 = f_2 / f_1$, $T_1 / T_0 = T_2 / T_1$ and $E_1 / E_0 = E_2 / E_1$. And the total redshift, pulse stretching, and energy fluence lost in the round trip of the pulse from A to B and back to its source (by time dilation and pulse stretching, but not by inertial drag) are therefore given by $f_2 = f_0 [i(\Delta x)]^{-2}$, $T_2 = T_0 [i(\Delta x)]^2$, and $E_2 = E_0 [i(\Delta x)]^{-2}$. The energy lost by the pulse in making a round trip over a short distance Δx is the same as the energy lost by travelling $2\Delta x$ in a straight path. Only the absolute distance travelled matters.

These simple scaling arguments for fractional energies apply only over a short distance $\Delta x \ll R$. In Ch. 6, the fractional energy dissipated through inertial drag will be shown to decay exponentially, rather than linearly.

In terms of photons, the scaling of power between A and B as $[i(\Delta x)]^{-3}$ can be understood as follows. In Fig. 4.2, suppose A radiates a plane-wave pulse of photons having a central frequency f_0, a pulse duration T_0, and an areal number density (number of photons per unit cross-sectional area) N_0. The central frequency of the photons measured at B must be redshifted to $f_1 = f_0 / i(\Delta x)$. Similarly, the pulse duration measured at B must be stretched to $T_1 = T_0 i(\Delta x)$. The dissipation of pulse energy in the inertial field is represented by the loss of photon areal number density as $N_1 = N_0 / i(\Delta x)$. The attenuation of photon areal number density from the pulse by doing work against the inertial drag force is irreversible. Thus we find the energy flux, also called intensity or power density, of a plane-wave electromagnetic pulse, after travelling a short distance Δx in an inertial field, is reduced by the factor

$$(N_1 f_1 / T_1) / (N_0 f_0 / T_0) = [i(\Delta x)]^{-3}. \tag{4.2}$$

Next, consider a distant light source at rest in the inertial field of a static, homogeneous universe. The wavelength of the redshifted light observed at the origin, $\lambda(0)$, is longer than the wavelength of the light emitted at x, $\lambda(x)$, by a factor $\lambda(0)/\lambda(x) = e^{-P(x)/2}$. The redshift parameter is defined as $Z(x) \equiv [\lambda(0)/\lambda(x)]-1$. From Eq. (4.1), the exact redshift parameter in a static homogeneous universe,

$$Z(x) = -1 + (\sinh C_0) / \sinh[C_0(1 - x/R)], \tag{4.3}$$

is plotted in Fig. 4.3 in the Cartesian coordinate x along any ray.

For all C_0 over short ranges, $x \ll R$, the redshift is linear with range, as

$$Z(x) \approx (C_0 / \tanh C_0) x / R. \tag{4.4}$$

Over short ranges, the Hubble constant, H_0, is related to the redshift parameter by $Z(x) \approx H_0 x / c$ for $Z \ll 1$, so that for $x \ll R$,

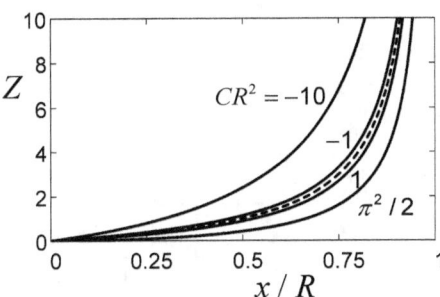

Fig. 4.3. Redshift parameter in static universe *vs.* normalized Cartesian coordinate x/R for normalized curvatures CR^2 indicated. Dashed curve is $C = 0$.

$$H_0 \approx (C_0 / \tanh C_0) c / R. \tag{4.5}$$

Experimental determinations of H_0 are summarized by [10], including the conclusion by the HST Key H_0 Group [11] that $H_0 = 71 \pm 6$ km$/$(s$\cdot$Mpc). More precise measurements of H_0 from the cosmic microwave background are model-dependent and do not supplant these earlier measurements [10]. From Eq. (4.5), Fig. 4.4 relates the normalized curvature CR^2 to the event horizon R for several values of H_0 spanning this uncertainty range.

For $|C|R^2 \ll 1$, the spacetime interval, Eq. (2.14), and the redshift parameter, $Z \approx x/(R-x)$, are independent of curvature. Over short ranges in this limit, $Z \approx x/R$, so that $R \approx c/H_0$. As shown in Fig. 4.5, therefore, the inertial field in a static, *nearly flat* universe produces an event horizon at a radius equal to the Hubble age of the universe (times c). The determination of H_0 by the HST Key H_0 Group [11], $H_0 = 71 \pm 6$ km$/$(s$\cdot$Mpc) in a *static universe* with $|C|R^2 \ll 1$, corresponds to an event horizon located at about $R = 13.8 \pm 1.2$ billion light years.

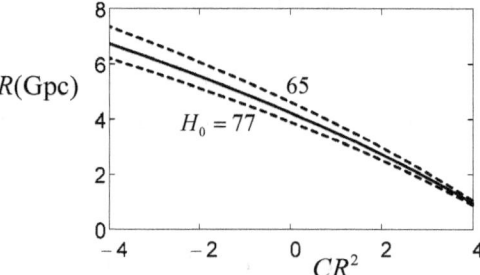

Fig. 4.4. Radius of static universe *vs.* normalized curvature CR^2 at 'uncertainty bounds' of Hubble constant, 65 to 77 km/(s$\cdot$Mpc) (dashed curves) and at 71 km/(s$\cdot$Mpc) (solid).

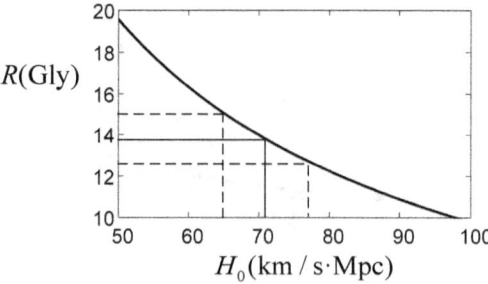

Fig. 4.5. Radius of static universe related to Hubble constant for $|C|R^2 \ll 1$. $H_0 = 71 \pm 6$ km/(s$\cdot$Mpc) implies $R = 13.8 \pm 1.2$ billion light years.

Chapter 5
Appearance of Cosmic 'Acceleration' in a Static Universe

This section calculates the departure from a linear cosmological redshift that occurs as a consequence of an inertial field acting on light propagating through a *static*, homogeneous universe. In a static universe, the effects on the redshift parameter of an inertial field can appear to have been caused by an acceleration of an expanding universe. The redshift caused by inertial drag in a *static universe* is shown here to be consistent with observations of the luminosity of Type Ia supernovas (SN-Ia). An actual acceleration of an expanding universe, on the other hand, appears inconsistent with the *cosmological principle* of homogeneity and with the *Copernican principle*.

Type Ia supernovas result from the thermonuclear explosion of white dwarf stars when their mass exceeds the Chandrasekhar limit of $1.44 M_\odot$, where $M_\odot = 1.99 \times 10^{30}$ kg is the mass of our Sun. Because SN-Ia share similar features, their intrinsic (absolute) luminosities are related, and they are regarded as 'standard candles' from which range measurements may be inferred.

About the same time, two groups reported the first detailed observations of apparent luminosity *vs.* redshift for SN-Ia, the Supernova Cosmology Project [12, 13] and the High-z Supernova Search Team [14, 15]. Since then, statistical uncertainties of observations of SN-Ia have been reduced, in large part by improving the corrections for light-curve shapes, which are the energy-flux histories measured at different wavelengths. The SN-Ia data are presented as apparent magnitude corrected for variations in absolute magnitude as indicated by light-curve shapes, according to several different models. The light-curve shapes are fitted to families of parametrized curves for SN-Ia. The various light-curve fitting models generally assume that SN-Ia are intrinsically the same at high and low redshift for a given shape and color. The light-curve fitting models calculate distances by fitting each light curve for peak magnitude, stretch parameter, and color.

From Ch. 4, as summarized in Table 5.1, the transmission of pulses of light from SN-Ia to our detectors is affected by: (i) inertial time dilation; (ii) pulse stretching; and (iii) inertial drag. Pulse stretching has no effect on pulse energy, but reduces pulse power irreversibly. Inertial drag causes a dissipative decay of pulse energy with a constant decay rate in a static, nearly flat universe proportional to the pulse energy, as $dE/dx = -E/R$.

Table 5.1. Energy and power transmission factors for light traveling over distance x in static, nearly flat universe. Heavy border indicates total power transmission factor used for distance modulus of SN-Ia in Fig. 5.1.

	Transmission Factor	
Effect	Energy	Power
Time Dilation	$1-x/R$	$1-x/R$
Pulse Stretching	1	$1-x/R$
Inertial Drag	$e^{-x/R}$	$e^{-x/R}$
Total	$(1-x/R)e^{-x/R}$	$(1-x/R)^2 e^{-x/R}$

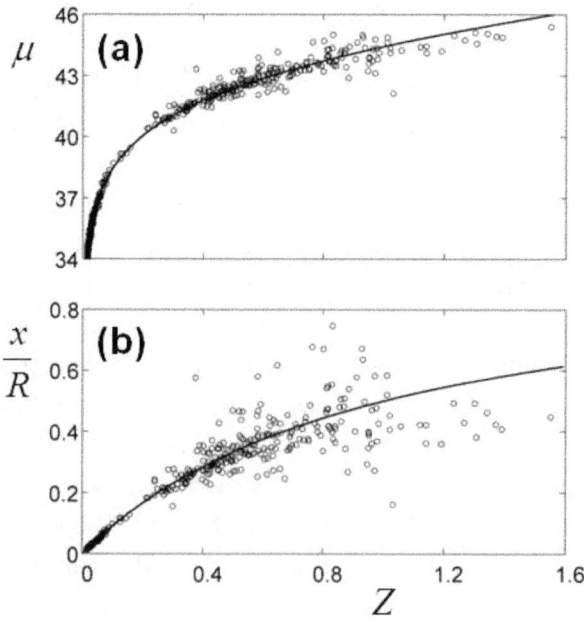

Fig. 5.1. Curves are: (a) distance modulus from Eq. (5.5); and (b) normalized range from Eq. (5.4) of SN-Ia vs. redshift z in nearly flat ($C_0 \ll 1$), static universe. Data from [16].

Let the absolute luminosity, or energy emitted per second, by each SN-Ia be L_0. The apparent luminosity l_1 is the SN-Ia radiation energy received per second per cm^2 of receiving area. In the inertial field of a flat-3-volume static universe, corresponding to the metric of Eq. (2.13), the ratio of absolute luminosity to apparent luminosity over a short distance Δx is, from Eq. (4.2),

$$L_0 / l_1 = 4\pi(\Delta x)^2 [i(\Delta x)]^3, \tag{5.1}$$

where $i(\Delta x)$ is given by Eq. (4.1). Of the three factors of $i(\Delta x)$ in Eq. (5.1), corresponding to Table 5.1, one is due to redshift by inertial time dilation, one is due to pulse stretching, and one is due to energy dissipation by work done against the inertial field of the mass in the universe. Energy dissipates from the electromagnetic pulse by attenuation through a process described in Ch. 4 that leads to a loss of fractional energy, rather than of energy itself, per unit distance. So the factor $i(\Delta x)$ in Eq. (5.1) due to energy dissipation by work done against the inertial field of the mass over a cosmologically significant distance x is $\exp[i(x)-1]$, rather than $i(x)$, and the ratio L_0 / l_1 over a long distance x is

$$L_0 / l_1 = 4\pi x^2 [i(x)]^2 \exp[i(x)-1], \tag{5.2}$$

corresponding to the transmission factor in the heavy-border box in Table 5.1. Since the redshift parameter $Z(x)$ in a static universe is related to the time-dilation factor $i(x)$ by $Z(x) = i(x)-1$, the ratio L_0 / l_1 over a long distance x is

$$L_0 / l_1 = 4\pi x^2 [Z(x)+1]^2 \exp[Z(x)]. \tag{5.3}$$

The apparent magnitude of an SN-Ia is $m_1 = -2.5 \log_{10} l_1 + \text{constant}$. The absolute magnitude M_0 of an SN-Ia is related to its absolute luminosity by $M_0 = -2.5 \log_{10} L_0 + \text{constant}$. Then the distance modulus, $\mu_1 \equiv m_1 - M_0$, defined as the difference between the apparent and absolute magnitudes, for an SN-Ia at a distance x in a static universe is

$$\mu = \mu_0 + 5\log_{10}[(x/R)(Z+1)e^{Z/2}], \tag{5.4}$$

where μ_0 is a constant. In a spatially flat ($C_0 = 0$), static universe, the exact redshift parameter given by Eq. (4.3) becomes $Z(x) = x/(R-x)$, and the distance modulus becomes

$$\mu(Z) = \mu_0 + 5\log_{10}(Ze^{Z/2}).$$

(5.5)

This distance modulus function of the redshift parameter Z is shown in Fig. 5.1(a) for the value of the instrument-calibration constant, $\mu_0 = 43.35$, that minimizes the sum of squares of deviations from Eq. (5.5). The 397 data points are the Constitution set of SN-Ia data with a SALT light-curve fitter [16]. Unlike other models that are fitted to data, this calculation of the distance modulus from Eq. (5.5) involves no adjustable parameters. The constant μ_0 depends only on the calibration of the detectors and not on the features of the model.

Displaying the SN-Ia data on a linear scale of range vs. Z, as in Fig. 5.1(b), rather than a logarithmic scale of distance modulus vs. Z, as in Fig. 5.1(a), gives a different impression of the scatter in the SN-Ia data. From Eq. (5.4), Fig. 5.1(b) plots the normalized range, $x/R = Z/(Z+1)$, in a spatially flat ($C_0 = 0$), static universe against Z with the same data from [16].

From Eq. (4.3), the exact redshift parameter in a static universe may be expanded in a Taylor series expansion about $x/R = 0$ as

$$Z(x) = \left(\frac{C_0}{\tanh C_0}\right)\frac{x}{R} + \left(\frac{C_0^2}{\sinh^2 C_0} + \frac{C_0^2}{2}\right)\frac{x^2}{R^2} + \dots .$$

(5.6)

If the Hubble constant is defined as $c\,dZ(x)/dx$ in the limit $x/R \to 0$, then $H_0 \equiv (C_0/\tanh C_0)c/R$ is the exact Hubble constant in a static, homogeneous universe, and the expansion of $Z(x)$ becomes

$$Z(x) = (H_0 x/c) + [1 - (\tanh^2 C_0)/2](H_0 x/c)^2 + \dots .$$

(5.7)

To second order in x, the redshift parameter is of the form $Z(x) = (H_0 x/c)(1 + a_0 x/c^2)$, where $a_0 \equiv [1 - (\tanh^2 C_0)/2]cH_0$ is an 'acceleration' constant, predicting a departure from a linear cosmological redshift at long ranges. In a spatially flat ($C_0 = 0$), static universe, the 'acceleration' constant becomes $a_0 = c^2/R$ and the redshift parameter becomes $Z(x) = (H_0 x/c)/(1 - x/R)$.

A spherically symmetric acceleration of an expanding universe is possible only if exceptions are made to the *cosmological principle* of homogeneity and to the *Copernican principle*. The *cosmological principle* is Einstein's idea that the universe is homogeneous in the large-scale average. But radial acceleration is only possible if there are radial gradients in density or pressure, that is, if there are radial inhomogeneities. The *Copernican principle* is the idea that we on Earth do not occupy a central, privileged position as observers of the universe. For example, the distribution of mass in the universe should not appear to be spherically symmetric only to us and not to observers anywhere else in the universe.

But mass in the universe does appear to be distributed about us with a high degree of spherical symmetry and a high degree of homogeneity. Observations are consistent with the absence of large-scale voids or the so-called Hubble bubble, a departure of the local Hubble constant from the average value [16-18].

Figure 5.2(a) shows that a universe that appears to us to expand about the origin with a velocity proportional to distance, that is, one that appears to us to be unaccelerated, also appears to any other observers at rest in their local frames to be unaccelerated. Figure 5.2(b), on the other hand, shows that a universe that appears to us to expand symmetrically about the origin with a velocity that increases faster (or slower) than linearly with distance appears to any other observers at rest in their local frames to be expanding *asymmetrically* about themselves. Thus, an accelerating universe does not appear to satisfy the *Copernican principle*.

Indeed, [19] found that an improved maximum-likelihood analysis based on a much larger database of supernovas is "still quite consistent with a constant rate of expansion," that is, no acceleration.

One way to distinguish between an accelerating universe and no cosmic acceleration is to measure cosmic acceleration directly. A method has recently been proposed by [20] for doing this by measuring the Sandage-Loeb [21, 22] drift in the redshift of the hydrogen absorption line at 21-cm over the course of a decade or so. If the results of direct measurements are consistent with the acceleration of a lambda-cold-dark-matter (ΛCDM) universe, then the hypothesis of a static, homogeneous universe, first proposed by Einstein, would be substantially confuted. On the other hand, if direct measurements of cosmic acceleration give a null result, then, as outlined in Appendix F, the Einstein static-universe hypothesis would warrant further testing.

26

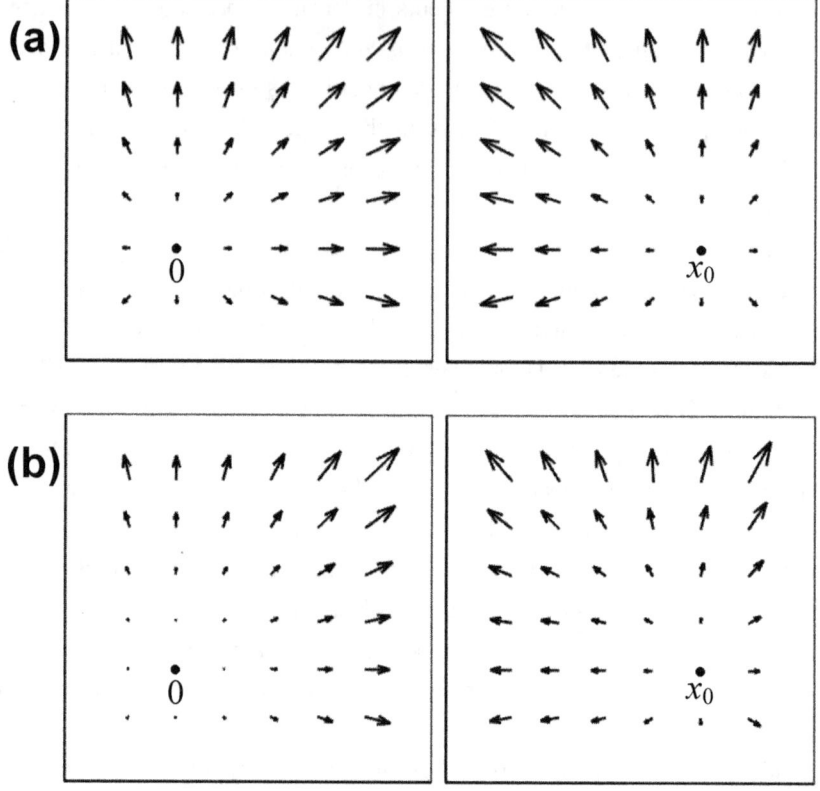

Fig. 5.2. Vector field plot showing appearance to observers at 0 (left) and x_0 (right) of universe expanding symmetrically about 0 with: (a) linear velocity increase; and (b) acceleration.

Chapter 6
Exact Particle Motion in a Static Inertial Field

This section calculates the exact motion of a particle in the inertial field of a static, homogeneous universe. The inertial field exerts a drag force on any particle moving with respect to the rest frame of a static universe. This drag force is proportional to a universal inertial drag constant, which is an invariant scalar.

The model of the universe for calculating inertial drag is of a static, homogeneous medium having no physical properties other than some constant linear combination of cosmological constant and energy density, which might include pressure. Such a universe has no privileged coordinate origin, but does have a privileged reference frame F, the frame at rest with respect to all the mass. Particle velocities may be defined absolutely with respect to F. In such a translationally symmetric universe, inertial drag does not depend on position or direction.

Symmetry requires a particle at rest in F to remain at rest. A particle at rest will be in a static equilibrium with the mass of the universe. An alternative description is that every particle is at the bottom of a potential well. If a particle is displaced from its rest position in F, it must modify its gravitational potential with the rest of the mass of the universe. To do that, it must radiate enough energy to modify its potential and to cause the mass of the universe to respond to the new potential accordingly. A particle at rest in F has no energy to radiate and therefore is 'locked' in position by the inertial field.

The energy radiated by a moving particle must be of a magnitude and of a form that will cause other mass to move in response to the changed potential. That is, it must be dipole inertial radiation. A solitary particle moving through a static universe is an example of an unbound quadrupole, and the dipole inertial radiation of unbound quadrupoles, derived in [9], is discussed in Ch. 11 of this book.

Symmetry requires the drag force to be collinear with the particle velocity. The instantaneous velocity vector of a particle in F provides the only unique direction for calculating inertial drag. By symmetry, therefore, any inertial forces experienced by a particle moving freely through a static universe must be directed along the direction of its velocity in F. Since the physical mechanism for an inertial drag force involves radiation of dipole power, conservation of energy requires the inertial drag force to be directed opposite to the direction of particle velocity, so the particle does work on the inertial field.

Simple arguments show that the inertial drag force acting on a particle moving short distances at slow speeds must be constant. A particle in motion in F must constantly radiate energy to modify its gravitational potential with the rest of the mass of the universe. Consider a particle that is displaced linearly a short distance Δx. The particle will have radiated an energy $E(\Delta x)$ in modifying its gravitational potential with the rest of the mass in a static universe. Next, consider the same displacement to be performed in two equal steps of $\Delta x / 2$ each. Since the two steps are equivalent, in each step the particle must radiate an equal amount of energy. And since performing a displacement Δx in one step is equivalent to performing it in any number of steps, the energy radiated during a small displacement at a nonrelativistic speed must be proportional to the displacement. That means the radiated power must be proportional to the speed, and the inertial drag force must be independent of velocity as well as position, at least to lowest order in velocity and displacement.

Besides the dissipative power loss caused by inertial drag, there are two other kinds of power loss of a particle bunch that depend on the position of the observer. In the limit of ultrarelativistic velocities, the three types of power loss for particle bunches in an inertial field correspond qualitatively and quantitatively to the three kinds of power loss for pulses of light in an inertial field. With respect to behavior in an inertial field, photons can not readily be distinguished as being ultrarelativistic particles or being massless particles. In showing that an inertial field can account for the appearance of cosmic acceleration in a static universe, Chs. 4 and 5 discussed these three kinds of power loss and summarized them in Table 5.1. The three kinds of power loss for particle bunches, which are discussed in turn in this section, are:

- Observer-dependent power loss by stretching of particle bunches;
- Observer-dependent energy loss by inertial time dilation;
- Dissipative energy loss by inertial drag

In a static, homogeneous universe, the privileged reference frame F, the rest frame of the universe, is the only one in which a particle at rest remains at rest. (One also expects that F is the only frame in which the cosmic microwave background has no dipole anisotropy.) In the following, the motion of particles will be calculated with respect to observers at rest in F. Let $\dot{s}^\mu = [c\dot{i}, \dot{\mathbf{s}}]$ be the velocity 4-vector of a particle with respect to F, where an overdot indicates differentiation with respect to proper time τ of the particle, that is, $\dot{s}^\mu = ds^\mu / d\tau$, $\dot{\mathbf{s}}$ is the instantaneous specific momentum of the particle in F at the location of the particle, and s is a coordinate that acts like an odometer, keeping track of the total distance traveled by the particle in F.

For an observer at rest in F, the equation of motion of a particle in a static inertial field is

$$\ddot{s}^\mu + \Gamma^\mu{}_{\alpha\beta} \dot{s}^\alpha \dot{s}^\beta = d^\mu , \tag{6.1}$$

where $d^\mu = [d^0, \mathbf{d}]$ is the inertial drag 4-vector. With the static, homogeneous spacetime interval of Eq. (2.1) and the Christoffel symbols of Eqs. (A.4), the scalar product of the 4-velocity with the equation of motion is

$$e^P c^2 \dot{i} \ddot{i} - e^Q \dot{\mathbf{s}} \cdot \ddot{\mathbf{s}} + e^P c^2 \dot{i}^2 \dot{\mathbf{s}} \cdot \nabla P / 2 - e^Q \dot{s}^2 \dot{\mathbf{s}} \cdot \nabla Q / 2 = \dot{s}_\alpha d^\alpha . \tag{6.2}$$

The first integral of Eq. (6.2) is

$$e^P c^2 \dot{i}^2 - e^Q \dot{s}^2 = c^2 + 2 \int (\dot{s}_\alpha d^\alpha) d\tau . \tag{6.3}$$

Since the invariant spacetime interval, Eq. (2.1), leads to the energy equation,

$$e^P c^2 \dot{i}^2 - e^Q \dot{s}^2 = c^2 , \tag{6.4}$$

the inertial drag 4-vector must be orthogonal to the 4-velocity, that is, $\dot{s}_\alpha d^\alpha = 0$.

The time component of the equation of motion, Eq. (6.1), is

$$\ddot{i} + (\dot{\mathbf{s}} \cdot \nabla P) \dot{i} = d^0 / c . \tag{6.5}$$

The space components of the equation of motion are

$$\ddot{\mathbf{s}} + e^{P-Q} c^2 \dot{i}^2 (\hat{\mathbf{s}} \cdot \nabla P) \hat{\mathbf{s}} / 2 + (\dot{\mathbf{s}} \cdot \nabla Q) \dot{\mathbf{s}} / 2 = \mathbf{d} , \tag{6.6}$$

where $\hat{\mathbf{s}}$ is a unit 3-vector in the direction of instantaneous particle velocity in F. Since $\dot{\mathbf{s}}$ is anti-parallel to $\mathbf{d}$, and $\dot{s} \geq 0$, then $\dot{\mathbf{s}} \cdot \mathbf{d} = -\dot{s}d \leq 0$. And since $\dot{s}_\alpha d^\alpha = 0$, then the time and space components of the inertial drag 4-vector are related by

$$d^0 = -\dot{s}de^{Q-P} / ci \le 0.$$
(6.7)

The time component, d^0, is a measure of power loss by the particle. For a particle at rest ($\dot{s} = 0$) in F, $d^0 = 0$.

If a universal inertial drag constant D is defined by the invariant scalar quantity $D \equiv (-d_\alpha d^\alpha)^{1/2}$, then the inertial drag 4-vector is

$$d^\mu = -D[e^{-(P-Q)/2}\dot{s} / c, \, e^{(P-Q)/2}i\hat{s}].$$
(6.8)

One means of determining the magnitude of the drag constant D is by the following analysis. From Eq. (2.13), the exact diagonal metric in natural Cartesian coordinates of the inertial field of a static, homogeneous universe along any ray from the origin is

$$g_{00} = e^P = \frac{\sinh^2[C_0(1-r/R)]}{\sinh^2 C_0}, \quad -g_{11} = -g_{22} = -g_{33} = e^Q = 1,$$
(6.9)

where $C_0 \equiv (|C|R^2/2)^{1/2}$ and C is the curvature scalar. In these coordinates, the equation of motion, Eq. (6.6), becomes

$$\ddot{s} = \frac{i \sinh C_0(1-r/R)}{\sinh C_0}\left[-D + \frac{c^2 i(\hat{s}\cdot\hat{r})C_0 \cosh C_0(1-r/R)}{R \sinh C_0}\right],$$
(6.10)

where $\hat{r}$ is a unit vector from the observer at rest at the origin to the instantaneous particle position. For a particle at rest at the origin, $r = 0$ and $i = 1$, and the equation of motion becomes

$$\ddot{s}(0) = -D + (\hat{s}\cdot\hat{r})(C_0 / \tanh C_0)c^2 / R.$$
(6.11)

The condition that a particle at rest at the origin should not spontaneously accelerate in any direction is

$$D = (C_0 / \tanh C_0)c^2 / R = cH_0,$$
(6.12)

where H_0 is the Hubble constant defined at Eq. (5.7). With the range of values of the Hubble constant given by [10], $H_0 = 71 \pm 6 \text{ km} / (\text{s} \cdot \text{Mpc})$, the universal inertial drag constant, an invariant scalar, is

$$D = 0.69 \pm 0.06 \ \text{nm} / \text{s}^2 \ . \tag{6.13}$$

In terms of a dimensionless energy,

$$\gamma \equiv e^P i \ , \tag{6.14}$$

normalized to particle rest energy, and a dimensionless momentum

$$p \equiv e^{Q/2} \dot{s} / c \ , \tag{6.15}$$

Eqs. (6.5) and (6.8) give the loss rate of particle energy as

$$\dot{\gamma} = -De^{P/2} p / c \ . \tag{6.16}$$

From Eq. (6.4), the relation between γ and p is

$$\gamma^2 = e^P (1 + p^2) \ . \tag{6.17}$$

Combining Eqs. (6.16) and (6.17) gives the exact fractional energy loss rate with respect to proper time,

$$\dot{\gamma} / \gamma = -(D / c) p / (1 + p^2)^{1/2} \ , \tag{6.18}$$

which is proportional to the Hubble constant, $H_0 = D / c$.

The remainder of this section will calculate the power loss transmission factors for the three kinds of power loss experienced by particle bunches in an inertial field: (1) Bunch stretching; (2) inertial time dilation; and (3) inertial drag. In each of the three cases, each kind of power loss will be treated independently of the other two. The calculations will use natural Cartesian coordinates in a spatially flat universe, for which $C_0 = 0$, $e^Q = 1$, $e^P = (1 - r / R)^2$, $D = c^2 / R$, and the rate of a clock at rest in F , measured by an observer at the origin, is $i = (1 - r / R)^{-1}$, from Eq. (4.1).

Case 1. Observer-Dependent Power Loss by Stretching of Particle Bunches

The duration of a bunch of particles appears to be stretched and its power, but not its energy, appears to be reduced by time-interval stretching effects of an inertial field. This energy-conserving stretching effect is calculated independently of the energy-loss effects of inertial time dilation and inertial drag in this case by artificially maintaining the bunch velocity at a constant 'absolute velocity'. In the rest frame F of a static universe, an 'absolute velocity' $\mathbf{v}_F$ of a particle may be defined as the velocity of the particle measured by an observer who is *both* at rest in F *and* at the exact location of the particle at the instant of measurement. By this definition, the absolute velocity of a particle is independent of inertial time dilation. In considering the absolute velocity of a particle, one may imagine a linear array of observers at rest in F along the trajectory of a particle, each observer equipped with a clock and a measuring stick, and each observer making his measurement of absolute velocity at the instant the particle passes by. Consider a variation of such a configuration shown in Fig. 6.1.

The configuration in Fig. 6.1 is used to calculate the effects of bunch stretching independently of inertial time dilation and inertial drag by artificially maintaining a constant absolute velocity of particles. Observers A and B are at rest in F. Observers C and D are passengers in spaceships, which are launched consecutively from the position of A and which travel at a constant absolute speed v_0 to B, an 'absolute distance' L away. Absolute distance is defined as the distance measured by an observer at rest in F. In natural Cartesian coordinates, all observers at rest in F agree on absolute distances. Observers A and B measure the absolute speed of the spaceships, $v_F = v_0$, at time of launch and time of collision, respectively. The dimensionless 'absolute energy', defined as the energy, normalized to rest energy, measured by an observer who is *both* at rest in F *and* at the exact location of the particle at the instant of measurement, is simply $\gamma_F \equiv (1 - v_F^2 / c^2)^{-1/2}$, independent of inertial time dilation.

Observers A and B measure the absolute energy of the spaceships, $\gamma_0 \equiv (1 - v_0^2 / c^2)^{-1/2}$, at time of launch and time of collision, respectively. The two spaceships containing observers C and D might represent the front end and back end, respectively, of a bunch of particles. A stretching of the time interval between the spaceships then represents a stretching of the particle bunch.

Observers C and D can perform experiments, such as dipole anisotropy measurements of the cosmic microwave background, to measure their speeds in F, and each maintains a constant absolute speed v_0 for the entire flight by providing propulsion as necessary to overcome inertial drag. Each observer, A, B, C, and D, has his own clock. Table 6.1 indicates the times shown on each of the observers' clocks at each of the four events depicted in Fig. 6.1.

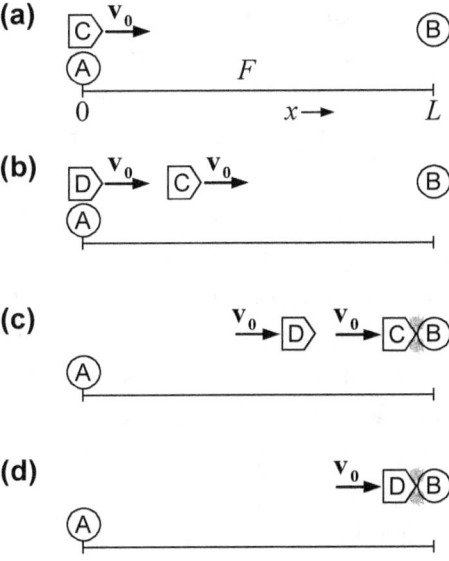

Fig. 6.1. A configuration involving four observers, A, B, C, and D, for deter-mining effects of bunch stretching alone, without inertial drag. A and B are at rest in F, and C and D maintain constant absolute velocity $\mathbf{v}_0$. Depicted are the: (a) launch of C; (b) launch of D; (c) collision of C with B; and (d) collision of D with B.

Table 6.1. Interval-stretching effects of inertial time dilation are seen by times of four events measured by clocks of observers, A, B, C, and D. The events correspond to those in Fig. 6.1(a) – (d).

	Launch of C	Launch of D	Collision of C with B	Collision of D with B
A	0	Δt_0	$t_A(L)$	$t_A(L) + \Delta t_0$
B	0	$\Delta t_0 / (1 - L/R)$	$t_B(L)$	$t_B(L) + \Delta t_0 / (1 - L/R)$
C	0	–	$L / \gamma_0 v_0$	–
D	0	Δt_0	–	$L / \gamma_0 v_0 + \Delta t_0$

As shown in Table 6.1, all observers, A, B, C, and D, synchronize their clocks to $t = 0$ at the moment of the launch of C. Observers A and D measure the launch of D an interval Δt_0 after the launch of C. By his own clock, observer B witnesses the launch of C at $t = 0$ and by inertial time dilation, witnesses the launch of D an interval $\Delta t_0 / (1 - L/R)$ later. To observer B, the interval between launches is stretched by the time-dilation factor $(1 - L/R)^{-1}$, just as a pulse of light from a supernova is stretched by this factor, as discussed in Chs. 4 and 5.

When C and D reach the position of B, they collide with B, as shown in Figs. 6.1(c) and 6.1(d). To observers C and D, the distance L is Lorentz-contracted by a factor γ_0, so the time interval from launch to collision according to their respective clocks is $\tau_C = \tau_D = L/(\gamma_0 v_0)$. Since the absolute speeds of C and D are the same, their flight times are the same. Then since by C's own clock, C collides with B at time $L/\gamma_0 v_0$, D, which was launched Δt_0 later, must collide with B at time $L/\gamma_0 v_0 + \Delta t_0$ by D's own clock, as indicated in Table 6.1.

Observer A sees the spaceships appear to decelerate through inertial time dilation from an initial speed of v_0 to a speed at collision of $v_0(1 - L/R)$. The apparent speed measured by A, $v_A(x) = v_0(1 - x/R)$, and apparent deceleration measured by A, $dv_A(x)/dt = -(1 - x/R)v_0^2/R$, during the flight of each spacecraft is a function of distance travelled. The flight time to reach x, measured by A during the flight of each spacecraft, must then be

$$t_A(x) = \int_0^x [v_0(1 - x/R)]^{-1} dx = -(R/v_0)\ln(1 - x/R), \qquad (6.19)$$

as indicated in Table 6.1.

Although observer B, unlike A, sees the spaceships appear to *accelerate* through inertial time dilation from an initial speed of $v_0(1 - L/R)$ to a speed at collision of v_0, the total flight time to reach B, measured by B, is the same as the flight time measured by A,

$$t_B(L) = t_A(L) = -(R/v_0)\ln(1 - L/R), \qquad (6.20)$$

as indicated in Table 6.1. But to observer B, D appeared to be launched a time interval $\Delta t_0 / (1 - L/R)$ later than C. And since C and D have equal flight times, observer B measures its time of collision with D as $-(R/v_0)\ln(1 - L/R) + \Delta t_0 / (1 - L/R)$, as indicated by the bold-outlined block in Table 6.1. That is, through interval-stretching by inertial time dilation, B *measures a longer time interval between collisions than* A *measures between launches*. The factor by which the collision interval is stretched relative to the launch interval, $(1 - L / R)^{-1}$, is the same as the factor by which light pulses from supernovas are stretched. For example, see Eq. (4.1) and Table 5.1.

Since the absolute velocity of each spacecraft is maintained constant during their flights, the absolute energy of each spacecraft at B is equal to the absolute energy of each spacecraft at A, γ_0. Next, we will see how these results change when absolute velocity is not maintained, and inertial time dilation causes the spacecraft to appear to lose energy.

Case 2. Observer-Dependent Energy Loss by Inertial Time Dilation

This case treats the energy lost by a particle as measured by an observer at rest in F. The effects of inertial drag are suppressed, so the only kind of power loss treated in this case is that caused by inertial time dilation. Unlike power loss caused by energy-conserving bunch stretching, inertial time dilation causes an actual loss of energy of particles between the source and the detector, but the apparent loss of energy is in general different for different observers.

For reasons explained in the next section, treating inertial time dilation without inertial drag is applicable to analysis of the Pioneer anomaly. Owing to the apparent high velocity of our solar system through F, of order 600 km/s, the Sun, the Earth, and the Pioneer spacecraft all experience an inertial drag force that is identical within a few percent, and is therefore undetectable at present. As will be shown in Ch. 7, inertial time dilation alone accounts for the nontransient component of the Pioneer anomaly measured from Earth.

In a spatially flat universe, the time and space components, Eqs. (6.5) and (6.6), in natural Cartesian coordinates become

$$\ddot{t} = -[1 - 2(\hat{s} \cdot \hat{r})\dot{t}]\dot{s} / (R - r),$$ (6.21)

$$\ddot{s} = -D\dot{t}(1 - r / R)[1 - (\hat{s} \cdot \hat{r})\dot{t}].$$ (6.22)

The first integral of Eq. (6.22) is the energy equation, Eq. (6.4). Eq. (6.21) can be integrated and combined with the energy equation to solve for the general motion of a particle, including effects of both inertial time dilation and inertial drag. Two special cases are of particular interest. The first special case, calculated here in Case 2, solves for the apparent energy loss of a radially inbound particle from inertial time dilation alone, and independently of inertial drag. The second special case, Case 3, solves for particle motion from inertial drag alone, and independently of inertial time dilation. Case 3 is of interest, for example, for analyzing the deceleration of stars and gas clouds within distant galaxies, as is done in Chs. 9 and 10.

To eliminate inertial drag artificially and calculate just the effects of inertial time dilation on a particle, the right-hand side of Eq. (6.1) is taken to be zero. Then for a radially inbound particle ($\hat{s} \cdot \hat{r} = -1$ and $\dot{r} < 0$), Eqs. (6.21) and (6.22) become

$$\ddot{t} = +2\dot{t}\dot{r} / (R - r) < 0, \tag{6.23}$$

$$\ddot{r} = +D(1 - r / R)\dot{t}^2 > 0. \tag{6.24}$$

The first integral of Eq. (6.23), for a radially inbound or a radially outbound particle, is

$$(1 - r / R)^2 \dot{t} = \gamma_0, \tag{6.25}$$

where the constant of the motion, $\gamma_0 \equiv (1 - v_0^2 / c^2)^{-1/2}$, is the dimensionless absolute energy of the particle at the origin. For an inbound or outbound particle slowed only by inertial time dilation and not by inertial drag, the energy equation, Eq. (6.17), becomes

$$(1 - r / R)(1 + \dot{r}^2 / c^2)^{1/2} = \gamma_0. \tag{6.26}$$

This energy equation for a radially inbound particle with no inertial drag is of particular interest in the limits of slow velocities and of ultrarelativistic velocities, like those of cosmic rays. In the limit of slow velocities ($\dot{r}^2 \ll c^2$), the energy equation, Eq. (6.26), for an inbound particle, stopped only by inertial time dilation, becomes

$$\dot{r}^2 \approx v_0^2 + 2Dr, \tag{6.27}$$

which is the equation of motion of a particle that has uniform deceleration $-D$.

In the limit of ultrarelativistic velocities ($\dot{r}^2 \gg c^2$), the energy equation for an inbound particle, slowed only by inertial time dilation, becomes

$$\dot{r} \approx -c\gamma_0 / (1 - r / R). \tag{6.28}$$

But for an ultrarelativistic particle, $\dot{r}$ is the specific momentum and $\dot{r}c$ is the specific energy of the particle. This result shows that inertial time dilation alone, without inertial drag, reduces the energy of a radially inbound ultrarelativistic particle by a factor $(1 - L/R)$, where L is the range from the observer to the source. This result is in agreement with the redshift of a light pulse in a static universe caused by inertial time dilation, as found in Ch. 5 and indicated in Table 5.1.

This result for change of energy by inertial time dilation is very much observer-dependent at low energies. The apparent change of energy for an out-bound particle is different from that just calculated for an inbound particle. At slow velocities, inertial time dilation causes an apparent uniform acceleration of an outbound particle by $+D$ (an acceleration that is offset by inertial drag). The energy loss from an inbound particle is not a dissipative loss. Instead, it is more like an observer-dependent conversion of kinetic energy to potential energy.

An analogy to the observer-dependent change of energy of a particle caused by inertial time dilation is the change of energy in a uniform gravitational field. Consider an observer at some height h above the ground in a uniform gravitational field of magnitude D. A particle thrown from the ground will lose specific energy Dh in rising to the observer. But a particle thrown downward by the observer will appear to this observer to accelerate downward and, if it bounces off the ground elastically, will appear to decelerate upward to return to the observer with the same energy. Where the analogy of a gravitational field to an inertial field breaks down is that every observer at rest in F will disagree with every other observer as to the effect of inertial time dilation on the motion of a given particle.

In trying to understand the meaning of energy loss by inertial time dilation, it may be helpful to consider that in a spatially flat universe, every slow particle climbs the wall of a 'potential well' with a constant gradient c^2/R. That is, the work done by a particle being virtually displaced from the origin to the radius R of the universe against the nondissipative 'time-dilation potential', as measured by a detector at the event horizon, is just the rest energy of the particle,

$$\text{Work} = \int_0^R (m_0 D) dr = m_0 c^2. \tag{6.29}$$

In a sense, inertial time dilation 'binds' every particle in a static universe with a binding energy equal to the rest energy of the particle.

Case 3. Dissipative Energy Loss by Inertial Drag

This case treats the dissipative energy loss experienced by a particle doing work against the inertial field. The effects of inertial time dilation are suppressed by considering the particle to move at a constant distance relative to an observer at the origin, a distance much larger than the limited range of the particle. This case is applicable to displacements of particles that are much smaller than the range to an observer, and over which differences in inertial time dilation are insignificant. Examples include orbits of stars and gas clouds within distant galaxies, treated in Chs. 8, 9, and 10. This case is also applicable to calculating the absolute velocity of a particle in F. Inertial drag is the only one of the three kinds of power loss mechanisms that is independent of the observer and depends only on distance travelled in F.

The configuration in Fig. 6.2 is used to calculate the effects of inertial drag on a particle independently of inertial time dilation, by maintaining a constant absolute distance L to the particle by an observer at rest in F. As shown in Fig. 6.2, the particle is assumed to bounce back and forth between two walls separated by a distance Δz much shorter than the range of the particle L to the observer at the origin. The collisions of the particle with the walls are assumed to be perfectly elastic, so that, according to the observer at the origin, the only energy lost by the particle is through inertial drag.

For a particle moving normal to the line of sight of an observer at the origin at a constant range L, $\hat{s} \cdot \hat{r} = 0$, and Eqs. (6.21) and (6.22) become

$$\ddot{t} = -\dot{s} / R\alpha_0, \tag{6.30}$$

$$\ddot{s} = -D\alpha_0 \dot{t}, \tag{6.31}$$

where $\alpha_0 \equiv 1 - L / R$ is a constant. The first integral of Eq. (6.30) is

$$\dot{t}(s) = \dot{t}(0) - s / R\alpha_0. \tag{6.32}$$

By the definition of γ in Eq. (6.14), the initial value of γ at $s = 0$ is $\gamma_0 = \alpha_0^2 \dot{t}(0)$. By substitution of Eq. (6.32), the exact equation of motion, Eq. (6.31), for a particle subject only to inertial drag, and not to inertial time dilation, becomes

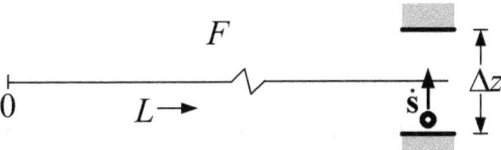

Fig. 6.2. A configuration for determining effects of inertial drag alone, without inertial time dilation. An observer at rest at the origin in F watches a particle at a fixed range L bouncing with elastic collisions between two walls separated by $\Delta z \ll L$.

$$\ddot{s} - (c/R)^2 s = -\gamma_0 D / \alpha_0, \tag{6.33}$$

But from the energy equation, Eq. (6.17),

$$\gamma_0 = \alpha_0 (1 + \dot{s}_0^2 / c^2)^{1/2}, \tag{6.34}$$

where $\dot{s}_0 \equiv \dot{s}(0)$ is the initial value of $\dot{s}$ at $s = 0$. Therefore, the equation of motion becomes

$$\ddot{s} - (c/R)^2 s = -D(1 + \dot{s}_0^2 / c^2)^{1/2}. \tag{6.35}$$

The first integral of this equation of motion gives

$$(1 + \dot{s}^2 / c^2)^{1/2} = (1 + \dot{s}_0^2 / c^2)^{1/2} - s/R. \tag{6.36}$$

Integrating again gives the exact distance travelled in F by a particle subject to inertial drag, as a function of the proper time of the particle,

$$s(\tau)/R = (\dot{s}_0 / c)\sinh(c\tau/R) - (1 + \dot{s}_0^2 / c^2)^{1/2}\left[\cosh(c\tau/R) - 1\right]. \tag{6.37}$$

As for Case 2, this equation of motion for a particle subject only to inertial drag is of particular interest in the limits of slow velocities and of ultrarelativistic velocities. For a particle subject only to inertial drag and not inertial time dilation, in the limit of slow velocities ($\dot{s}_0 \ll c$), Eqs. (6.36) and (6.37) become, respectively,

$$\dot{s}^2 \approx \dot{s}_0^2 - 2Ds, \tag{6.38}$$

$$s(\tau) \approx \dot{s}_0 \tau - D\tau^2 / 2.$$ (6.39)

Until the particle stops, these equations describe the motion of a nonrelativistic particle that has uniform deceleration $-D$. The range in F of a slow particle subject only to inertial drag is $\dot{s}^2 / 2D$. This is the same range as that given by Eq. (6.27) for an inbound particle stopped only by inertial time dilation and not inertial drag.

The inertial drag 4-force acting on a particle of rest mass m_0 is defined as $f^\mu \equiv m_0 d^\mu$. From Eq. (6.8), then, the inertial drag 3-force is

$$\mathbf{f} = -m_0 \alpha_0 D\dot{s}\,\hat{\mathbf{s}},$$ (6.40)

and the rate of change of the absolute energy $E_a \equiv m_0 c^2 \varepsilon_a(r)$ of the particle due to its work against the inertial drag force is

$$\dot{E}_a = \mathbf{f} \cdot \dot{\mathbf{s}} = -m_0 \alpha_0 D\dot{s}\dot{s},$$ (6.41)

where $\varepsilon_a(r) \equiv (1 + \dot{r}^2 / c^2)^{1/2}$ is the dimensionless absolute energy.

In the limit of ultrarelativistic velocities ($\dot{s}_0 \gg c$), Eq. (6.32) becomes

$$\dot{t} \approx \dot{s}_0 / c\alpha_0,$$ (6.42)

for a particle slowed only by inertial drag. Combining Eqs. (6.41) and (6.42) gives

$$dE_a / ds \approx -m_0 D\dot{s}_0 / c.$$ (6.43)

But for an ultrarelativistic particle, the dimensionless momentum $\dot{s}/c$ at each step in distance ds is about equal to the dimensionless absolute energy, ε_a, of the particle. Then the dissipative energy loss per unit distance traveled in F is

$$dE_a / ds \approx -E_a / R.$$ (6.44)

This result shows that inertial drag alone, without inertial time dilation, reduces the momentum and the energy of an ultrarelativistic particle by a factor

$$p(s) / p(0) \approx E_a(s) / E_a(0) \approx \exp(-s / R).$$ (6.45)

This result is in agreement with the dissipative energy loss of a light pulse in a static universe caused by inertial drag, as found in Ch. 5 and indicated in Table 5.1. This dissipative energy loss by inertial drag is independent of the observer and is evinced by an attenuation of energy from the light pulse, rather than by a redshift of frequency. In a particle model of a light pulse, the number of photons in the pulse is attenuated by the factor $\exp(-s/R)$.

Case 4. Total Energy Loss from Inbound Particle by Inertial Time Dilation and Inertial Drag

This case combines the effects of inertial time dilation and inertial drag on a particle radially inbound towards an observer at the origin. The time and space components of the equation of motion for an inbound particle ($\hat{s} \cdot \hat{r} = -1$ and $\dot{r} = -\dot{s} < 0$) are

$$\ddot{t} = +(2\dot{t}+1)\dot{r}/(R-r) < 0,$$ (6.46)

$$\ddot{r} - D(1-r/R)\dot{t}^2 = +D(1-r/R)\dot{t} > 0.$$ (6.47)

The first integral of Eq. (6.46) is

$$(1-r/R)^2(\dot{t}+1/2) = \varepsilon_a(0) + 1/2,$$ (6.48)

where the constant of the motion, $\varepsilon_a(0)$, is the dimensionless absolute energy of the particle at the origin.

From Eqs. (6.6) and (6.8), the inertial drag 3-force is

$$\mathbf{f} = -m_0 D(1-r/R)\dot{t}\,\hat{s},$$ (6.49)

The rate of change of the absolute energy E_a of the inbound particle due to its work against the inertial drag force is

$$\dot{E}_a = \mathbf{f} \cdot \dot{s} = -m_0 D(1-r/R)\dot{t}\dot{s} < 0.$$ (6.50)

Then the energy equation, Eq. (6.4),

$$(1-r/R)\dot{t} = \varepsilon_a(r),$$ (6.51)

is used to eliminate $\dot{t}$ from Eqs. (6.47) and (6.50), giving

$$\ddot{r} - D\varepsilon_a^{\ 2} / (1 - r / R) = + D\varepsilon_a,$$ (6.52)

$$\dot{\varepsilon}_a = -\varepsilon_a \dot{s} / R.$$ (6.53)

As discussed in the introduction of this section, the work done by a particle against the inertial drag force results in energy lost to dipole inertial radiation. The dipole inertial radiation of unbound quadrupoles, such as particles moving freely through the universe, is calculated in Ch. 11. Just as the equation of motion of a charged particle does not account for the reactive effects of dipole Larmor radiation from the particle, so too does the equation of motion of a particle in an inertial field, Eq. (6.52), not account for the reactive effects of energy dissipation by dipole inertial radiation. And just as the Abraham-Lorentz modified equation of motion addresses this shortcoming for charged particles, so too can Eq. (6.52) be modified in an approximate and time-averaged way to include the reactive effects of energy dissipation by inertial drag.

The first integral of Eq. (6.52) is

$$\frac{d\varepsilon_a}{dr} - \frac{\varepsilon_a}{R - r} = \frac{1}{R}.$$ (6.54)

According to Eq. (6.53), however, the constant term $1 / R$ on the right-hand side of Eq. (6.54) should be replaced by ε_a / R to account for the reactive effects of dissipative energy loss at high velocities. This result suggests that the dissipative energy loss per unit distance is proportional to the relativistic mass of the particle, $m_0 \varepsilon_a$, not the rest mass, m_0. This result is not unexpected, because the energy radiated by a moving particle must be of a magnitude and of a form that will cause other mass in the universe to move in response to the changing potential of the particle, and the changing potential of the particle is proportional to its relativistic mass, not its rest mass. The modified equation of motion for an inbound particle, including the reactive effects of energy dissipation, becomes

$$\frac{d\varepsilon_a}{dr} - \frac{\varepsilon_a}{R - r} = \frac{\varepsilon_a}{R}.$$ (6.55)

The solution, shown in Fig. 6.3, relates the absolute energy of a particle at a detector, E_{det}, to its absolute energy at its source, E_{src}, at range r,

$$E_{det} = E_{src}(1-r/R)\exp(-r/R).$$ (6.56)

This solution for absolute energy in an inertial field applies equally to all particles, slow and relativistic. This solution determines characteristic features of the
cosmic ray spectrum. It even applies to massless particles like photons. Equation (6.56) is in agreement with the total energy loss of a light pulse in a static
universe caused by the combined effects of inertial time dilation and inertial
drag, as found in Ch. 5 and indicated in Table 5.1. The energy loss by inertial
time dilation causes a redshift of the frequency of the light by a factor $(1-r/R)$
. The dissipative energy loss by inertial drag causes an attenuation of energy
from the pulse by a factor $\exp(-r/R)$, rather than a frequency redshift. From
Eq. (6.52) or Eq. (6.56), the deceleration of a slow ($\varepsilon_a \approx 1$) inbound particle,
caused in equal parts by the combined effects of inertial time dilation and inertial drag, is $\ddot{s} \approx -2D$.

From Cases 3 and 4, the absolute energy of a particle after travelling a distance s in F, and being affected only by observer-independent inertial drag, is

$$\gamma_F(s) = \left[1 - v_F^2(s)/c^2\right]^{-1/2} = \gamma_0 \exp(-s/R).$$ (6.57)

From Eq. (6.57), Fig. 6.4 shows the normalized absolute speed v_F/c of particles as a function of normalized distance s/R travelled in F for several values of normalized initial absolute speed v_0/c. The dotted curve shows the initial absolute speed of a particle that results in a particle having a range s in F.
Equation (6.57) shows that particles with an initial normalized specific energy
$\gamma_0 > e^1$, that is, a normalized initial absolute speed $v_0/c > (1-e^1)^{1/2} = 0.795$,
have a range $x_0 > R$.

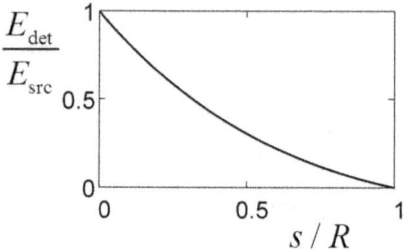

Fig. 6.3. Ratio of energy of inbound particle measured at a detector to energy at source *vs.* normalized distance travelled by particle in spatially flat ($C_0 = 1$) static universe. Particle stops at range for which $E_{det} = m_0 c^2$.

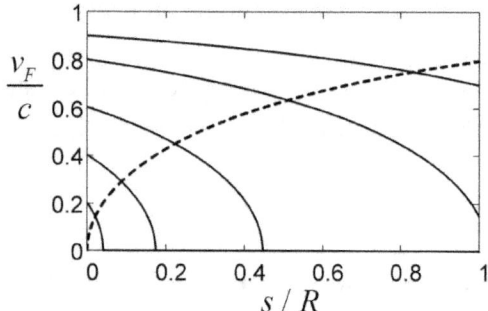

Fig. 6.4. Normalized absolute speed *vs.* normalized distance travelled by particles for several values (solid curves) of initial absolute speed in spatially flat ($C_0 = 0$) static universe. Dashed curve is normalized initial absolute speed v_0 / c *vs.* normalized range x_0 / R.

Chapter 7
Inertial Time Dilation and the Pioneer Anomaly

This section shows how time dilation in the inertial field of a static, homogeneous universe can account for the Pioneer anomaly. The Pioneer anomaly is what appears to be a nearly constant acceleration of magnitude 0.874 ± 0.133 nm/s^2 directed towards the Sun or Earth that was measured on both the Pioneer 10 and Pioneer 11 spacecraft after they reached a range of about 15 to 20 A.U. from Earth [23, 24].

Many authors have proposed explanations, reviewed in [24], for the apparent anomalous acceleration of the Pioneer spacecraft. Some explanations involve gravitational mechanisms, and some are associated with $c^2 / R \approx 1$ nm/s^2. Some have emphasized that any explanation of the Pioneer anomaly must be consistent with the precisely measured orbital precession rates of inner and outer planets (see, *e.g.*, [25, 26]). The issue of consistency with planetary precession rates is addressed in Ch. 8, which shows that inertial drag has no appreciable effect.

Consider a nearly flat ($C_0 \ll 1$ and $D \approx c^2 / R$) static universe with the metric of Eq. (2.13). In the slow-velocity approximation, the dimensionless coordinate speed, $\beta \equiv v / c \ll 1$, is about equal to the dimensionless momentum, $p \equiv \dot{s} / c \approx \beta$. In this same slow-velocity, short-range approximation ($r \ll R$), the dimensionless energy defined in Eq. (6.14) becomes

$$\gamma \approx (1 - 2r / R)\dot{i} . \tag{7.1}$$

But in this slow-velocity, short-range approximation, Eq. (6.17) becomes

$$\gamma \approx 1 - r / R + \beta^2 / 2 , \tag{7.2}$$

so that

$$\dot{i} - 1 \approx r / R + \beta^2 / 2 . \tag{7.3}$$

The first term on the right-hand side of Eq. (7.3) is the inertial time dilation term, and the second term is the relativistic time dilation term.

As was found in Ch. 6, the deceleration of a slow particle by inertial drag is

$$dv/dt \approx -D, \tag{7.4}$$

where the universal drag constant from Eq. (6.13) is $D = cH_0 = 0.69 \pm 0.06 \text{ nm}/\text{s}^2$. Then from Eqs. (7.2) and (7.4), the apparent rate of loss of energy of a particle of rest mass m_0, measured at the origin, for a slow particle moving radially from the origin in a static, spatially flat $(C_0 = 0)$ universe, is

$$d(\gamma mc^2)/dt \approx -2mDv, \tag{7.5}$$

half of which is due to observer-dependent inertial time dilation and half to dissipative inertial drag. As noted in Ch. 6, under these conditions of slow velocities and short ranges, the rate of loss of dimensionless energy, $d\gamma/dt \approx -2v/R$, is proportional to the speed.

These equations of motion in the slow-velocity approximation apply only to measurements made by an observer *at rest* in the universal rest frame F of a static universe, and not to an observer in motion with respect to F. Our Sun, however, seems to be moving through the local rest frame of our universe much faster than even the orbital speeds of its planets about the Sun. Motion through the local rest frame is inferred by a dipolar anisotropy in the cosmic microwave background. A slight dipolar temperature anisotropy measured by the WMAP satellite experiment [27] seems to indicate that the local group of galaxies is moving with a speed of 627 ± 22 km/s with respect to the cosmic microwave background [28, 29].

If our Sun were moving with respect to F at more than 600 km/s, then in F, the difference Δv in the velocity vectors of the Sun v_0 and the Pioneer spacecraft v_p would be nearly negligible. That is because the velocities Δv relative to the Sun are only about 12.2 km/s for Pioneer 10 and 11.6 km/s for Pioneer 11 [23]. Since the magnitude of the inertial drag force is independent of velocity in the slow-velocity approximation, the inertial drag acting on the Sun and on the Pioneer spacecraft would differ only slightly in direction, and virtually not at all in magnitude. That is, the inertial drag deceleration acting on the Sun would be nearly identical in magnitude and direction to the inertial drag deceleration acting on both Pioneer spacecraft.

As illustrated schematically in Fig. 7.1, in the rest frame F, the velocity of the Sun $\mathbf{v}_{\odot}$ differs by only $\Delta \mathbf{v} = \mathbf{v}_P - \mathbf{v}_{\odot}$ from the velocity of the Pioneer spacecraft $\mathbf{v}_P$, where if $v_{\odot} \approx 600\,\text{km/s}$, then $\Delta v \approx 12\,\text{km/s}$ is about $0.02 v_{\odot}$. In the slow-velocity approximation applicable here, the inertial drag acting on the Sun, $d\mathbf{v}_{\odot}/dt$, differs only slightly from the drag acting on the Pioneer spacecraft, $d\mathbf{v}_P/dt$. As seen in Fig. 7.1, the difference in magnitude between the inertial drag acting on the Sun and on the Pioneer spacecraft is about $(\Delta v / v_{\odot}) D \sin \psi$, an acceleration that depends on direction, but does not exceed about $0.02D$.

If the Sun is indeed moving faster than about 600 km/s with respect to F, then *in the rest frame of the Sun*, inertial drag on the Pioneer spacecraft should be negligible, and the only cause in this analysis that could account for the apparent anomalous slowing of the spacecraft in a nearly flat, static universe is inertial time dilation with an observer-dependent 'potential' Dr, increasing with distance r from the observer. In the slow-velocity approximation, therefore, at ranges much greater than 1 A.U., inertial time dilation results in an apparent acceleration, $a_p \approx D$, of the Pioneer spacecraft, directed towards the observer of each of the Pioneer spacecraft.

Figure 7.2 compares the predicted apparent acceleration from inertial time dilation, $a_p \approx 0.69 \pm 0.06\,\text{nm/s}^2$ of the Pioneer spacecraft at ranges beyond about 15 A.U. with measurements from [23] that report an anomalous acceleration of $a_p = 0.874 \pm 0.133\,\text{nm/s}^2$. In Fig. 7.2, error bars of ± 15.2 percent were arbitrarily applied to each data point, corresponding to the overall uncertainty in the measurements of [23]. In this analysis, the Sun's gravitational field is $g_{\odot} \equiv GM_{\odot}/r^2$.

If the cause of the anomalous acceleration is inertial time dilation, as measured from Earth, then the apparent acceleration should be Earth-pointing, with annual variations in direction as the Earth orbits the Sun. At distances greater than about 10 A.U., the annual variation measured from Earth of the inertial time dilation at the Pioneer spacecraft becomes smaller than the standard deviation of the measurements. Indeed, a careful analysis of recently recovered Doppler data for both spacecraft in 2011 [30], found "no support in favor of a Sun-pointing direction over the Earth-pointing" direction.

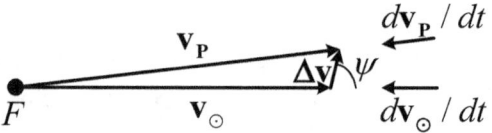

Fig. 7.1. Inertial drag on Pioneer spacecraft differs by less than about $0.02\,D$ from inertial drag D on Sun.

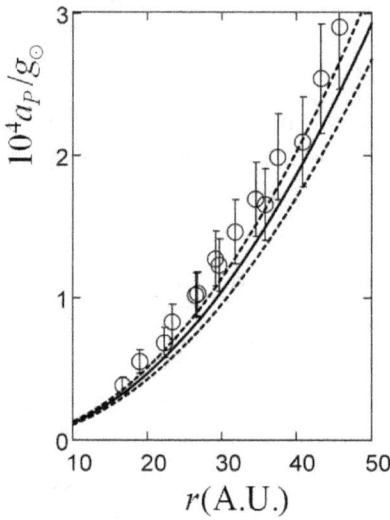

Fig. 7.2. Apparent deceleration a_p caused by inertial time dilation and normalized to Sun's field $g_\odot$ vs. range (in A.U.) for $D = 0.69\,\text{nm}/\text{s}^2$ (solid curve) $\pm 0.06\,\text{nm}/\text{s}^2$ (dashed curves). Pioneer 10 and 11 data (circles) from [23]. Error bars of ± 15.2 percent applied to each data point.

This analysis also found that the data favor a slow decay of the anomalous acceleration at a rate of about 0.017 nm/s²/yr [30]. The decaying acceleration is attributed to anisotropic emission of thermal radiation off the spacecraft [31]. Even though the uncertainties in the measurements in Fig. 7.2 and in the value of the drag constant D suggest inertial time dilation could be wholly responsible for the Pioneer anomaly, the anomalous acceleration does seem to be systematically greater than D in a way that would suggest another cause, such as thermal radiation, might be contributing. That is, the anomalous acceleration, a_p, might comprise a constant component D plus a component $\tilde{a}_p(T_L)$, which decays as a function of time after launch, T_L.

The best linear fit of the model, $a_p(T_L) = D + \tilde{a}_p(T_L)$, to the combined Pioneer data in [30] and for $D = 0.69 \, \text{nm/s}^2$ is

$$\tilde{a}_p(T_L) = [0.42 - (0.017 \, / \, \text{yr})T_L] \, \text{nm/s}^2, \tag{7.6}$$

as displayed in Fig. 7.3.

The next section will show that the inertial field has no discernible effect on planetary orbital motion about our Sun beyond that of the gravitational field of the Sun.

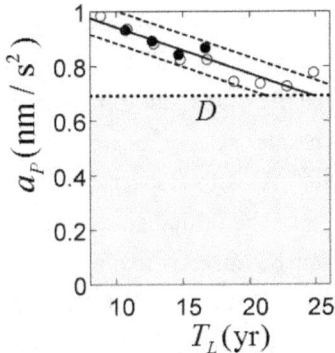

Fig. 7.3. Anomalous acceleration *vs.* years after launch of Pioneer 10 (open circles) and Pioneer 11 (filled circles) from [30]. Best linear fit of decaying acceleration $\tilde{a}_p$ from [30] added to constant apparent acceleration caused by time dilation (shaded), for $D = 0.69 \, \text{nm} \, / \, \text{s}^2$ (solid) $\pm 0.06 \, \text{nm} \, / \, \text{s}^2$ (dashed).

Chapter 8
Effects of Weak Inertial Drag on Orbital Motion

This section calculates the effects on orbital motion of the inertial field of a static, homogeneous universe. An important conclusion of this section is that, although weak inertial drag does cause the orbital radius, angular momentum, and energy to decay, it does not cause any appreciable precession of the periapsis of the orbit.

If the central mass, m_0, of a planetary system is at rest in the universal rest frame F of a static universe, then the exact metric of the combined static Schwarzschild gravitational field and static inertial field is given by Eq. (3.4). More pertinent to our own solar system, however, is the case of a central mass drifting at constant velocity $\mathbf{V}_0$ with respect to F with a drift speed much less than c, but much greater than the orbital speeds of the planets about the central mass. This case is relevant because, as explained in Ch. 7, our local group of galaxies seems to be drifting with a speed of 627 ± 22 km/s with respect to the cosmic microwave background [28, 29]. By symmetry, a cosmic microwave background should be at rest with respect to F in a static universe.

This section and the supporting calculations in Appendix B calculate the effects of weak inertial drag on a particle moving in a gravitationally bound orbit about a mass m_0 in an otherwise static, homogeneous universe. The calculations treat the effects of the inertial field on planetary motion as very small perturbations to the motion in the absence of an inertial field. The calculations are presented in detail in Appendix B for two cases: (1) An orbit about a mass m_0 at rest in the rest frame F of the static universe; and (2) an orbit at an orbital speed much slower than the drift velocity of the mass m_0 in F.

Case 1. An Orbit about a Mass at Rest in F

Using the exact diagonal spacetime metric of a mass m_0 at the origin of a static, homogeneous universe given by Eq. (3.4) and the exact equation of motion of the orbiting particle in a static inertial field from Ch. 6, the inertial drag 4-vector is given by Eq. (6.8) as

$$d^\mu = -D\left[W\dot{s}/c,\; W^{-1}i\hat{\mathbf{s}}\right], \tag{8.1}$$

where $\hat{\mathbf{s}}$ is a unit 3-vector in the direction of instantaneous particle velocity $\dot{\mathbf{s}} \equiv \mathbf{u}$, s is a coordinate measure of the total distance traveled by the particle in F, and

$$W(r,x) \equiv \left(1+\frac{r_s}{r}\right)^3 \left(\frac{\sinh[C_0(1-x/R)]}{\sinh C_0} - \frac{r_s}{r}\right)^{-1}. \tag{8.2}$$

In the weak Newtonian gravitational field of a mass m_0, terms of higher order than $\dot{s}^2/c^2$ and Gm_0/rc^2 are neglected, and there is no difference between the specific momentum of an orbiting particle, $\mathbf{u} = d\mathbf{s}/d\tau$, and its coordinate velocity, $\mathbf{v} = d\mathbf{s}/dt$. For the range x of the observer to the orbiting particle satisfying $r \ll x \ll R$, inertial time dilation and its variation over the orbit is negligible to the observer. In this modified-Newtonian approximation that neglects relativistic effects but allows for inertial drag, the energy equation from Eqs. (B.1) and (B.2) becomes

$$c^2 = (c^2 - 2Gm_0/r)\dot{t}^2 - [\dot{r}^2 + r^2\dot{\theta}^2 + (r\sin\theta)^2\dot{\phi}^2], \tag{8.3}$$

and the time component of the equation of motion becomes

$$c^2\ddot{t} + 2(Gm_0/r^2)\dot{r}\dot{t} = -2D\dot{s}. \tag{8.4}$$

Appendix B derives the Newtonian orbital energy equation, modified by dissipative inertial drag, as

$$E(s) = -Gm_0/r + \dot{s}^2/2 = E_0 - 2Ds, \tag{8.5}$$

where $E_0 \equiv -Gm_0/r_0 + \dot{s}_0^2/2 < 0$ is the initial (at $s = 0$) Newtonian specific energy $E(s)$ at the initial particle speed $\dot{s}_0$ and radius r_0, and derives the equation of motion as

$$\ddot{\mathbf{s}} + (Gm_0/r^2)\hat{\mathbf{r}} = -2D\hat{\mathbf{s}}. \tag{8.6}$$

In terms of the specific angular momentum of the orbiting particle, $l = r^2\dot{\phi}$, the equation of motion becomes

$$\ddot{r} - l^2/r^3 + Gm_0/r^2 = -2D(dr/ds),$$ (8.7)

$$\dot{l} = -2D(r^2 d\phi/ds).$$ (8.8)

If $D \ll Gm_0/r^2$ and $D \ll v^2/r$, then the effects of weak inertial drag on orbital dynamics can be calculated by the simple perturbation analysis detailed in Appendix B. The zeroth-order elliptical particle orbit for $D = 0$ is $r = a_0(1 - \varepsilon_0^2)/(1 + \varepsilon_0 \cos\phi)$, where a_0 is the constant semi-major axis and ε_0 is the constant eccentricity of the ellipse. The constant energy of the particle for $D = 0$ is $E_0 = -Gm_0/2a_0$, independent of eccentricity and angular momentum. And the constant specific angular momentum l_0 is given by $l_0^2 = 2a_0^2(1 - \varepsilon_0^2)|E_0| = Gm_0 a_0(1 - \varepsilon_0^2)$. The first-order perturbation analysis in D, described in Appendix B, gives the changes of energy, angular momentum, semi-major radius, and eccentricity over an orbital path length Δs for $D\Delta s \ll Gm_0/r$ and $D\Delta s \ll v^2$.

The decay of orbital energy with Δs to first order is

$$\Delta E/|E_0| = -2D\Delta s/|E_0|.$$ (8.9)

The decay of semi-major radius with Δs to first order is

$$\Delta a/a_0 = -2D\Delta s/|E_0|.$$ (8.10)

The fractional rate of decay of angular momentum,

$$\dot{l}/l_0 = -2D/v,$$ (8.11)

depends on eccentricity. To first order in $D\Delta s/|E_0|$, the decay of angular momentum with Δs,

$$\frac{\Delta l}{l_0} = \frac{-D\Delta s}{|E_0|}\left(\frac{1 - \varepsilon_0^2}{1 + 2\varepsilon_0 \cos\phi + \varepsilon_0^2}\right),$$ (8.12)

is shown in Fig. B.3 of Appendix B. Appendix B shows that the growth of eccentricity with Δs, to first order and for small eccentricity, is

$$\Delta\varepsilon/\varepsilon_0 \approx +D\Delta s/2|E_0|.$$ (8.13)

Case 2. An Orbit Drifting through F Much Faster than Orbital Speed

In this case, m_0 is moving through F at a constant velocity $\mathbf{V}_0$ much faster than the orbital velocity $\mathbf{v}$ of a particle about m_0. That is, $v \ll V_0 \ll c$. The gravitational field is strong enough that $D \ll Gm_0/r^2$ and $D \ll v^2/r$, and a first-order perturbation analysis in $D\Delta s / |E_0|$ is appropriate. Appendix B shows that the equation of motion in the rest frame of the mass m_0 is

$$dv / dt + Gm_0\mathbf{r}/r^3 = (D/V_0)\hat{\mathbf{V}}_0 \times [\hat{\mathbf{V}}_0 \times \mathbf{v}(t)]. \qquad (8.14)$$

In terms of the specific angular momentum of the orbiting particle, $l = r^2\dot{\phi}$, the equation of motion becomes

$$\ddot{r} - l^2/r^3 + Gm_0/r^2 = -(D/V_0)[\dot{r} - v_0 \sin\theta_0 \cos(\phi - \phi_0)], \qquad (8.15)$$

$$\dot{l} = -(D/V_0)[l_0 + rv_0 \sin\theta_0 \sin(\phi - \phi_0)], \qquad (8.16)$$

where $v_0 \equiv \hat{\mathbf{V}}_0 \cdot \mathbf{v}(t)$. The first-order perturbation analysis in D gives the small changes of energy, angular momentum, semi-major radius, and eccentricity over a time $t \ll V_0/D$. For a solar system moving through F with a speed $V_0 = 600$ km/s, this approximation is valid for times much shorter than 10 million years.

As calculated in Appendix B, the angular momentum of a planet decays with time as

$$l(t) = l_0(1 - Dt/V_0), \qquad (8.17)$$

and the approximate solution of the orbit is

$$r = a(t)[1 - \varepsilon^2(t)]/[1 + \varepsilon(t)\cos\phi], \qquad (8.18)$$

for $a(t) = a_0(1 - 2Dt/V_0)$ and $\varepsilon(t) = \varepsilon_0(1 - Dt/V_0)$. Since $E(t) = -Gm_0/2a(t)$ the orbital energy decays as $E(t) = -(Gm_0/2a_0)(1 + 2Dt/V_0)$.

Figure 8.1 compares this approximate solution, Eq. (8.18), to the exact numerical solution of Eqs. (8.15) and (8.16) for $\varepsilon_0 = 0.5$, $DT_0/2V_0 = 0.01$, $\theta_0 = \pi/4$, and $\phi_0 = \pi/4$, where $T_0 \equiv 2\pi(a_0^3/Gm_0)^{1/2}$ is the orbital period. A notable feature of the solution is that inertial drag does not cause the periapsis to precess in the orbital plane. Inertial drag, therefore, could make no appreciable

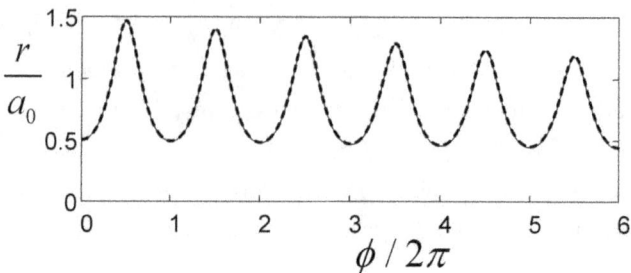

Fig. 8.1. Distance from m_0, normalized to a_0, vs. azimuthal angle of particle orbiting m_0 at speed $v \ll V_0$ with $\varepsilon_0 = 0.5$. Solid curve is numerical solution of Eqs. (8.15) and (8.16). Dashed is analytic solution, Eq. (8.18).

contribution to the precession rate of planetary orbits, such as the precession of Mercury's perihelion by 43 seconds of arc per century that is caused by general relativistic effects. Also notable is that the effects of inertial drag in the orbital plane are largely independent of the orientation of the orbital plane with respect to $\mathbf{V_0}$.

Chapter 9
Flat Rotation Curves and Tully-Fisher Relation in Clusters, Galaxies, and Gases

This section shows that Newtonian dynamics, without resorting to dark matter models, can account for the flattening of rotation curves at the outer edges of spiral galaxies and for the Tully-Fisher relation between the mass and velocity of pressure-supported systems, including spiral galaxies.

The Newtonian gravitational field outside a spherical distribution of mass m_0 at radius r is about Gm_0 / r^2. In the absence of inertial drag, a particle with speed v is maintained in a circular orbit at radius r by a centripetal acceleration, v^2 / r. Outside a spherical mass, therefore, Newtonian dynamics suggests the speed of a particle in a nearly circular orbit should be roughly $v = (Gm_0 / r)^{1/2}$.

The rotation curve of a galaxy is a measurement of the orbital speed $v(r)$ of gas as a function of distance r from the center of the galaxy. In such calculations, Newton's shell theorem, which applies to spherically symmetric mass distributions, is generally invoked to suggest that only mass within a radius r in the equatorial plane contributes to the gravitational force at r. Newton's shell theorem states that the field outside a spherically symmetric mass is the same as though all the mass is concentrated at the center, and the field inside a spherically symmetric shell is zero.

Using the approximation of circular orbits in a spherically symmetric galaxy, therefore, the orbital speed is given by $v^2 = Gm(r)/r$, where $m(r)$ is the mass contained within r. Beyond the visible rims of the galaxies, where $m(r)$ is expected to be constant, rotation curves are expected to fall as $v(r) \sim r^{-1/2}$ in this approximation. Rather than falling roughly as $r^{-1/2}$ as expected from Newtonian dynamics, however, orbital velocities of hydrogen gas clouds appear to flatten with distance from the center, that is, to be roughly constant at some rotational speed v_r, instead of falling in accordance with Newtonian dynamics beyond R_{rim}, which is the rim of the visible mass of the galaxy. See, for example, [32, 33].

Some rotation curves have been observed to fall more slowly or to be flat at and beyond the visible rims of the galaxies. That has been taken to indicate the presence of invisible matter beyond the rims. Observations of flattening of rotation curves have led to the now widely accepted concept that there must be mass beyond the visible edge of spiral galaxies, mass in a form that is undetectable by all of our sensors and instruments. According to this concept, this dark matter is in a form that is readily detectable only by its gravitational influence. With the proper density distribution, a 'halo' of this invisible dark mass could certainly account for the higher than expected rotational speeds beyond the visible edge of spiral galaxies. This dark matter could comprise black holes, cold white dwarfs, brown dwarfs, or clouds of elementary or exotic particles that might surround the galaxies, as reviewed for example in [33].

Besides accounting for the flattening of rotation curves outside spiral galaxies, this concept of dark mass has an additional attractive feature for cosmologists. The now standard lambda cold dark matter (ΛCDM) model of the universe and of cosmological structure formation depends on this cold dark matter contributing several times as much to the energy density of the universe as familiar baryonic matter, and depends on dark energy contributing many times as much. The widely accepted concordance model predicts that the energy density of the universe is roughly 70% dark energy, 25% cold dark matter, and only 5% baryonic matter. Such a high proportion of undetectable components would not be a problem for the dark matter concept, except that the properties of dark matter in galaxies seem not to be completely consistent with the properties on the cosmological scale.

For example, much observational evidence of galactic emissions in the reliable near-infrared band supports the Tully-Fisher relation [34] that the luminosity of spiral galaxies is proportional to v_r^4, where v_r is the asymptotic flat rotation speed. Without elaborate modifications, the standard ΛCDM model, on the other hand, requires a luminosity proportional to v_r^3 [33]. Besides failing to produce the observed Tully-Fisher relation, the ΛCDM model has other problems with observations on the galactic scale. The model "fails to predict the form of the rotation curves of low-mass galaxies; it leaves unexplained the systematic differences between the rotation curves of [low surface brightness] LSB and [high surface brightness] HSB galaxies; it predicts the existence of many unseen small companion satellite galaxies ..." [33]. Of course, the most acutely felt issue is that dark matter has not yet been detected.

More successful and consistent with observations on the galactic scale than the cold dark matter concept perhaps has been the modified Newtonian dynamics (MOND) model, first proposed by [35], and recently reviewed in [36] and [37] and [38]. The MOND model supposes that there exists a threshold constant with dimensions of acceleration, $a_0 \approx 0.1 \, \text{nm/s}^2$, below which the Newtonian gravitational acceleration g_N is modified, in that the effective gravitational acceleration approaches $(g_N a_0)^{1/2}$ below this threshold [39].

MOND has been successful in accounting for phenomena on scales ranging from dwarf galaxies to superclusters [39]. The predictions of MOND are consistent with flattening of rotation curves of spiral galaxies, and unlike dark-matter concepts, directly consistent with the Tully-Fisher relation. MOND also seems to account for other observations that are not predicted by dark-matter concepts. For example, the surface density of spiral galaxies seems to be limited to the order of a_0 / G [40], while dark matter does not suggest any such limit. The radius of nearly isothermal, pressure-supported systems of mass m, from globular clusters to clusters of galaxies, seems to be limited to the order of $(Gm / a_0)^{1/2}$ [41], while nothing in dark-matter concepts suggests any limit for such systems.

The MOND concept is less successful than dark matter, however, at explaining phenomena at the cosmological scale. And neither MOND nor dark matter explains phenomena at the scale of our solar system, like the Pioneer anomaly, because phenomena at the cosmological scale and the scale of our solar system seem to be related to an acceleration of the order of $D = 0.69 \pm 0.06 \, \text{nm/s}^2$, rather than $a_0 \approx 0.1 \, \text{nm/s}^2$, as seen for example in Chs. 4 through 8. This section and the next will show that the applicability of the inertial field in accounting indirectly for the flat rotation curves and Tully-Fisher relation in clusters, galaxies, and gas clouds is comparable to its applicability in accounting for phenomena at the cosmological and solar-system scales.

Pressure-supported systems, loosely bound by gravity, are modeled well by classical equations of hydrostatic equilibrium. Such systems include globular clusters within galaxies, elliptical galaxies, and clusters of galaxies, as well as gas clouds. Because spiral galaxies so clearly appear to be rotating owing to their pinwheel shapes, they are considered by many to be rotation-supported and not pressure-supported systems.

Chapter 10 will show that the appearance of rotation of stars about the center of spiral galaxies may often be misleading. In the outer regions of spiral galaxies, inertial drag will be shown to stop stars from rotating about the galactic center over a fraction of one orbit. To the extent that the outer regions of spiral galaxies are supported against gravity at all, therefore, they are generally pressure-supported, rather than rotation-supported.

The data in rotation-curve plots of velocity *vs.* radius beyond the visible disks of spiral galaxies are generated by observations of the 21-cm lines (at 1.42 GHz) of neutral hydrogen in *gases*, and not by observations of visible starlight. The actual rotation velocities of stars beyond the visible disk do not necessarily correspond to the measured velocities of gases there. Chapter 10 will show how gas clouds moving at high velocities relative to galaxies in an inertial field can be captured into bound orbits by the galaxies. Chapter 10 will also show that hot gas clouds are slowed less by inertial drag than cooler systems.

Because stars in spiral galaxies, particularly in the outer regions, may generally not be rotating about the centers of the galaxies, there is no need to invoke dark matter to balance the centrifugal force of a rotation in hydrostatic equilibrium. Instead, a simple model of clusters and galaxies and even spiral galaxies as pressure-supported isothermal spheres works well to account for flat rotation curves in the outer regions of spiral galaxies, and to account for other effects that would otherwise seem anomalous in spiral galaxies that are rotating.

The simple model of pressure-supported isothermal spheres in hydrostatic equilibrium was first calculated by [42]. The model features a mass density that falls as r^{-2} and a self-consistent gravitational field that falls as r^{-1} in the outer regions. The rotation curves in the outer regions of such pressure-supported isothermal spheres are therefore flat.

Globular clusters are spherical collections of stars generally comprising hundreds of thousands of old stars with a mean interstellar distance of order 1 light year. In our Milky Way galaxy, there are thought to be about 150 to 200 globular clusters. Elliptical galaxies are typically spheroidal collections of old stars supported against self-gravity by the random motion of the stars, with little evidence of galaxy rotation or recent star formation. Elliptical galaxies tend to be surrounded by many globular clusters. Clusters of galaxies seem to be the largest gravitationally bound objects in the universe, comprising hundreds to thousands of galaxies. The total mass of galaxy clusters is thought to be of the order of 10^{14} to 10^{15} solar masses, with about ten times as much mass in the hot, x-ray emitting gas between the galaxies as in the stars. A typical radius of a cluster of galaxies is 0.1 to 2 Mpc, and a typical spread of velocities for individual galaxies within the cluster is 800 to 1000 km/s.

For the purpose of determining the effects of an inertial field on pressure-supported loosely-bound galaxies, clusters, and gases, a simple model of such systems is used, having the following features. In this model, the system:

• is spherically symmetric about its center of mass;

• comprises a great number N of identical particles moving randomly in the center-of-mass frame, each particle having mass m and obeying Newton's laws of motion;

• has a drift velocity of its center of mass through the rest frame F of the universe of $\mathbf{V} = V(t)\hat{\mathbf{x}}$ in the x direction, affected only by inertial drag;

• is closed, except that energy is dissipated from the system through inertial drag;

• is characterized as an ideal gas with an isothermal temperature T in energy units, a mass density $\rho(r)$, and a pressure $\rho(r)T/m$;

• is characterized as a collisional gas, with a mean free path much shorter than the density scale length, and with negligible energy lost or time spent in elastic collisions;

• is subject to no forces other than the gravity of other particles and inertial drag, except during a collision;

• is loosely bound by gravity, meaning the interparticle distance is much greater than the size of the particles.

Within an isothermal sphere, the density and gravitational field may be modeled by the equation of static equilibrium,

$$d\rho/dr = m\rho(r)g(r)/T,\tag{9.1}$$

where, by Newton's shell theorem, the gravitational field at r is

$$g(r) = -GM(r)/r^2,\tag{9.2}$$

and

$$M(r) \equiv \int_0^r \rho(r)4\pi r^2 dr\tag{9.3}$$

is the mass of the system contained within radius r. These equations combine to give the second-order equation for $y(x) \equiv \ln[\rho_0 / \rho(r)]$,

$$\frac{d^2 y}{dx^2} + \frac{2}{x}\frac{dy}{dx} = \exp(-y), \tag{9.4}$$

where ρ_0 is the density at the center of the sphere, and $x \equiv r/L$ is the radius normalized to the scale length $L \equiv (T/4\pi Gm\rho_0)^{1/2}$. The initial conditions at $x = 0$ are $y(0) = 0$ and $dy(0)/dx = 0$. From the solution of Eq. (9.4), the gravitational field is given by $g(r) = -(T/mL)dy/dx$.

Figure 9.1 shows the solution of Eq. (9.4). The gravitational field is normalized to the peak field magnitude, $g_m = 0.517T/mL$. The radius of the peak field is $r_m = 3.01L$. The peak field and its radius are related by $g_m = 2.16(G\rho_0)r_m$. The density at r_m is $\rho(r_m) = 0.344\rho_0$. The mass contained within the radius r_m is $M(r_m) = 0.172r_m T/Gm$ or $M(r_m) = 0.239\rho_0 r_m^3$. The mass contained within the radius L is $M(L) = 0.586M(r_m)$.

An important feature of this solution is that density does not fall exponentially at large radius, as it does for an isothermal atmosphere in a uniform gravitational field. Instead, the density falls only as r^{-2}. More specifically, the solutions of Eqs. (9.1) through (9.3) at large radius, that is, at $r/L \gg 1$, are

$$\rho(r) \approx 2\rho_0 L^2 / r^2, \tag{9.5}$$

$$M(r) \approx (2T/Gm)r + M_0, \tag{9.6}$$

$$g(r) \approx -(2T/mr) - (GM_0/r^2), \tag{9.7}$$

where M_0 is a constant mass at the center.

Figure 9.2(a) compares the exact solution for density from Eq. (9.4) with the r^{-2} scaling of density for $r \gg L$ from Eq. (9.5). Figure 9.2(b) compares the exact solution for gravitational field from Eq. (9.4) with the r^{-1} scaling of gravitational field for $r \gg L$ from Eq. (9.7) for $M_0 = 0$.

According to Eq. (9.7), the gravitational field of a pressure-supported system scales as r^{-1} for $r \gg L$. This r^{-1} scaling of the gravitational field leads to flat rotation curves of pressure-supported gases and clusters for $r \gg L$. The 'rotation velocity' is defined as the speed $v_r \equiv [-g(r)\cdot r]^{1/2}$ at which the centripetal gravitational acceleration $g(r)$ is exactly balanced by the centrifugal

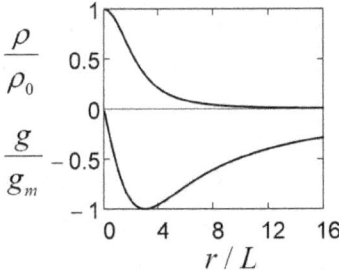

Fig. 9.1. Normalized density (upper curve) and normalized gravitational field (lower curve) *vs.* normalized radius of isothermal pressure-supported galaxies, clusters, and gases.

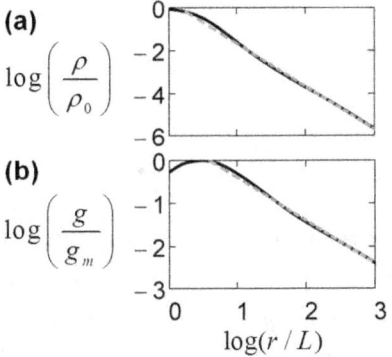

Fig. 9.2. (a) Normalized density (solid curve) scales as r^{-2} (dashed line) from Eq. (9.5), and (b) normalized gravitational field (solid curve) scales as r^{-1} (dashed line) from Eq. (9.7) for $r \gg L$ in pressure-supported systems.

acceleration v_r^2 / r for a particle in a circular orbit at radius r. From Eq. (9.7), the rotation velocity at large radius is expanded as

$$v_r(r) = \left(\frac{2T}{m}\right)^{1/2}\left[1 + \frac{GmM_0}{4Tr} + O\left(\frac{L}{r}\right)^2 + ...\right]. \tag{9.8}$$

The second term in Eq. (9.7), the expansion of the gravitational field at large radius, is the familiar Newtonian gravitational field of a mass M_0, which falls as r^{-2} outside the mass. The leading term in the expansion, $-2T/mr$, which dominates at large radius in isothermal pressure-supported systems, falls more slowly as r^{-1}.

Suppose the system has a high-density spherical core or bulge that is characterized only by a radius b and a mass M_0, and not by a temperature. This high-density core might represent, for example, an active galactic nucleus. Many galaxies show clear evidence for a central spheroidal component in the light distribution, and by inference in the mass of spiral galaxies. Because a bulge might have a different light-to-mass ratio than the disk, the potential presence of a bulge introduces one or two additional free parameters for fitting dark-matter models to rotation-curve data.

The presence of a central core or bulge changes the rotation curve in a pressure-supported galaxy. Let the density of an isothermal galaxy at the boundary of the bulge, $r = b$, be ρ_b. Since the galaxy outside the bulge is assumed isothermal, the equations of static equilibrium for $r > b$ are

$$d\rho / dr = m\rho(r)g(r) / T ,\tag{9.9}$$

$$g(r) = -\frac{G}{r^2}\left[M_0 + \int_b^r \rho(r)4\pi r^2\,dr\right].\tag{9.10}$$

These equations combine to give the same second-order equation for $y(x)$ as Eq. (9.4), but with the new normalizations, $y(x) \equiv \ln[\rho_b / \rho(r)]$ and $x \equiv r / L_b$, where $L_b \equiv (T / 4\pi Gm\rho_b)^{1/2}$ is the new scale length. The initial conditions at $x = b / L_b$ are $y(b / L_b) = 0$ and $dy(b / L_b) / dx = B$, where $B \equiv M_0 / 4\pi b^2 L_b \rho_b$. The gravitational field is $g(r) = -(T / mL_b)dy / dx$ for $r > b$. The gravitational field at $r = b$ is $g(b) = -GM_0 / b^2$.

Figure 9.3 shows a solution of the hydrostatic equilibrium equations, Eqs. (9.9) and (9.10), in the galaxy surrounding a central core of mass M_0 and radius b. The density is normalized to the density of the galaxy at the core boundary b, $\rho_b \equiv \rho(b / L_b)$. The gravitational field is normalized to T / mL_b. The particular solution in this figure is for a central core with mass and radius given by $GM_0 / b = 0.2T / m$ and $b = 0.1L_b$.

Figure 9.4(a) shows the normalized rotation curve for a pressure-supported system with no central mass M_0. In accordance with Eq. (9.8), the rotation curve at large radius is nearly flat at $v_r = (2T / m)^{1/2}$. Figure 9.4(c) shows the normalized rotation curve for a pressure-supported system with a central core characterized by $B = 1$ and $b = 0.6L_b$. The mass of the central core, therefore, is given by $GM_0 / b = 0.6T / m$. In this model, the constant mass density of the

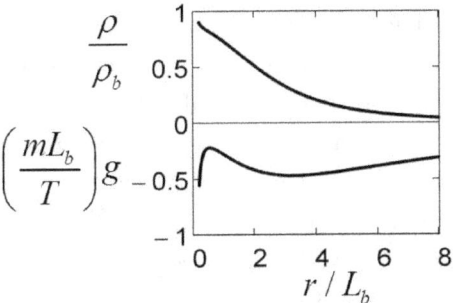

Fig. 9.3. Normalized density (upper curve) and normalized gravitational field (lower curve) *vs.* normalized radius of pressure-supported systems with central core, $GM_0 / b = 0.2T / m$ and $b = 0.1L_b$.

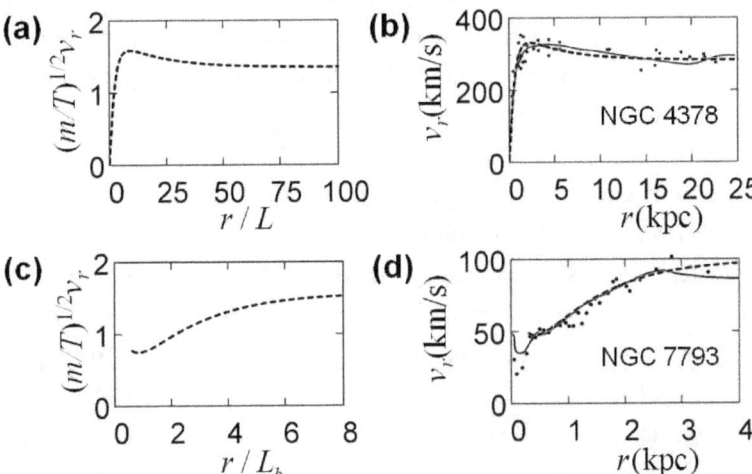

Fig. 9.4. Normalized rotation curves for pressure-supported systems with: (a) no central core; and (c) central core, $B = 1$ and $b = 0.6L_b$. Best fits (solid curves) by [43] to rotation curve data (points) by [43] for spiral galaxies: (b) NGC 4378 and (d) NGC 7793. Dashed curves in (b) and (d) are best fits of rotation curves in (a) and (c), respectively, to data by [43].

core is $\rho_{core} \equiv M_0 / (4\pi b^3 / 3)$. In the particular example shown in Fig. 9.4(c), $\rho_{core} = 1.0 \times 10^{-17} \, \text{kg/m}^3$. Just outside the core at $r = b$, the density is $\rho_b = 0.2 \times 10^{-17} \, \text{kg/m}^3$, that is, $\rho_b = 0.20\rho_{core}$. Rotation curves of pressure-supported systems at large radius are nearly flat at $v_r = (2T / m)^{1/2}$, whether or not the systems have a high-density core.

Figures 9.4(b) and 9.4(d) present the best fits (solid curves) to rotation curve data (points) by [43] measured in the visible disks of one of the larger spiral galaxies, NGC 4378, and one of the smaller spiral galaxies, NGC 7793, respectively. The rotation curves were fitted by [43] by assuming the mass is in a thin disk and the mass distribution is exactly proportional to the light, with mass-to-light ratios of 6.5 and 2.9, respectively.

When the rotation curve data in Fig. 9.4(b) is compared with the rotation curve of a pressure-supported system in Fig. 9.4(a), the following inferences may be drawn regarding a possible correspondence of a pressure-supported system with NGC 4378: $(T/m)^{1/2} \approx 210$ km/s, corresponding to $T/m \approx 44$ GJ/kg, or about ten thousand times the specific energy of TNT; $L \approx 0.25$ kpc; and $\rho_0 \approx 10^{-18}$ kg/m^3. When the rotation curve data in Fig. 9.4(d) is compared with the rotation curve of a pressure-supported system with a central core shown in Fig. 9.4(c), the following inferences may be drawn regarding a possible correspondence with NGC 7793: $(T/m)^{1/2} \approx 64$ km/s, corresponding to $T/m \approx 4.0$ GJ/kg; $L_b \approx 0.50$ kpc; $b \approx 0.30$ kpc; $M_0 \approx 1.8 \times 10^8 M_\odot$.

Central masses are not uncommon in galaxies. For example, supermassive black holes with millions to billions times the solar mass are believed to reside at the center of almost every galaxy. If NGC 7793 does have a central mass of order 10^8 solar masses in a spherical core of radius 0.3 kpc, that mass corresponds to a mass density just outside the core of order $\rho_b = 2 \times 10^{-18}$ kg/m^3. The central mass density of the much bigger and hotter galaxy, NGC 4378, is about $\rho_0 \approx 1 \times 10^{-18}$ kg/m^3. The scale length of NGC 4378, $L \approx 0.25$ kpc, is also comparable to the core radius of NGC 7793, $b \approx 0.30$ kpc. Freeman [40] observed that the central surface brightness of spiral galaxies was nearly constant from galaxy to galaxy. If central surface brightness is related to central mass density, as is generally supposed, then that might explain the comparable central mass densities of two such very different galaxies as NGC 4378 and NGC 7793.

As presented above, the model of an isothermal sphere has the unphysical property that the mass increases linearly with radius to infinity. In reality, an isothermal sphere must be truncated at a radius at which particles with speeds above a characteristic thermal speed can escape the gravitational field to infinity. Let R_{rim} be the radius at the rim of a pressure-supported isothermal sphere beyond which a significant number of stars in random thermal motion escape to infinity. From Eq. (9.7), the gravitational field at R_{rim} is

$$g_{rim} \approx -2T / mR_{rim}.$$ (9.11)

From Eq. (9.6), the mass of the isothermal sphere within R_{rim} is

$$M_{rim} \approx (2T / Gm)R_{rim}.$$ (9.12)

Combining Eqs. (9.11) and (9.12), and using the flat rotation velocity at large radius, $v_r = (2T / m)^{1/2}$, gives the relation between the velocity at the rim and the mass of an isothermal sphere as

$$M_{rim} \approx v_r^4 / G|g_{rim}|.$$ (9.13)

From Eq. (9.7), the gravitational potential of the truncated isothermal sphere for $r / L \gg 1$ is

$$\Phi(r) \approx \begin{cases} -v_r^2 \left[1 + \ln(R_{rim} / r)\right], & \text{for } r \leq R_{rim} \\ -v_r^2 (R_{rim} / r), & \text{for } r \geq R_{rim} \end{cases}.$$ (9.14)

The gravitational field of the truncated isothermal sphere for $r / L \gg 1$ is

$$g(r) \approx \begin{cases} -v_r^2 / r, & \text{for } r \leq R_{rim} \\ -v_r^2 (R_{rim} / r^2), & \text{for } r \geq R_{rim} \end{cases}.$$ (9.15)

The escape velocity for a particle moving radially outwards in the gravitational field of a truncated isothermal sphere for $r / L \gg 1$ is

$$v_{esc}(r) = [-\Phi(r)]^{1/2}.$$ (9.16)

Figure 9.5 shows the escape velocity and potential in the field of the truncated isothermal sphere for $r / L \gg 1$. Stars at the rim of a pressure-supported spiral galaxy or other pressure-supported system, moving outwards with a speed exceeding the rotation velocity, $v_r = (2T / m)^{1/2}$, can escape to infinity.

From Eq. (9.14), one finds that near $r = R_{rim}$, whether $r < R_{rim}$ or $r > R_{rim}$,

$$\Phi(r) \approx -v_r^2 (R_{rim} / r).$$ (9.17)

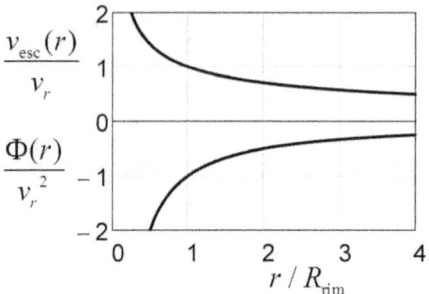

Fig. 9.5. Normalized escape velocity (upper curve) and normalized gravitational potential (lower curve) *vs.* normalized distance from center of truncated isothermal sphere for $r / L \gg 1$.

But the gravitational field is related to the potential by the defining equation,

$$g(r) = -d\Phi(r) / dr .$$ (9.18)

The only way that Eqs. (9.17) and (9.18) can both be satisfied near $r = R_{\text{rim}}$ is if R_{rim} is a constant independent of r, and dependent only on the isothermal temperature. That means the size of a truncated isothermal sphere is related to its temperature as

$$R_{\text{rim}} = 2T / ma_0 ,$$ (9.19)

where $a_0 \equiv |g_{\text{rim}}|$ is an acceleration constant, independent of r for all pressure-supported isothermal spheres. And the size of a truncated isothermal sphere is related to its rotation velocity at large radius as

$$R_{\text{rim}} = v_r^{\ 2} / a_0 .$$ (9.20)

From Eq. (9.12), the 'mean surface density' of a spiral galaxy, defined as $\sigma_m \equiv M_{\text{rim}} / \pi R_{\text{rim}}^{\ 2}$, is also a constant independent of r for all pressure-supported isothermal spheres,

$$\sigma_m = a_0 / \pi G .$$ (9.21)

For a_0 of order $0.06 - 0.1$ nm/s^2, the mean surface density of spiral galaxies modeled as pressure-supported truncated isothermal spheres is of order $0.3 - 0.5$

kg/m^2. This mean surface density corresponds roughly to an upper limit observed for the mean surface brightness of spiral galaxies, known as Freeman's law [40, 44]. This upper limit on mean surface brightness is observed primarily for high-surface-brightness (HSB) spiral galaxies.   Low-surface-brightness (LSB) galaxies appear to have lower mean surface density than the upper limit of Freeman's law, perhaps because they are dimmer and their brightness is not as easily measured, or perhaps because a_0 has some weak dependence on the temperature of spiral galaxies.

From Eq. (9.13), the mass of an isothermal sphere is related to its rotation velocity at large radius by

$$M_{rim} \approx v_r^4 / Ga_0.$$ (9.22)

Equation (9.22) is an expression of the Tully-Fisher relation [34]. Their work showed that the luminosity of a galaxy is proportional to the fourth power of the width of the 21-cm hydrogen line, suggesting that the mass of a galaxy is proportional to the fourth power of the rms velocity and of the rotation velocity. Sanders [33, 39] showed that the Tully-Fisher relation is applicable to pressure-supported systems ranging over six orders of magnitude in size, from globular clusters and massive molecular clouds in galaxies of 1 to a few tens of pc, and dwarf spheroidal galaxies and elliptical galaxies of 100 to a few tens of kpc, and clusters of galaxies of a few tens of kpc to a few Mpc. In all these pressure-supported systems, observations show that the acceleration constant a_0 in Eq. (9.22) is of the order of 0.1 nm/s^2, seemingly independent of temperature as measured by dispersion velocity. This comparison did not include spiral galaxies.

Since spiral galaxies will be shown in the next section to be pressure-supported in general, the size of spiral galaxies is also related to a_0 by Eqs. (9.19) and (9.20). Using data from [32], Table 9.1 shows the rotation velocity v_r measured at the outer edge of the visible portion, the rim radius R_{rim}, of each of several galaxies. From the data in Table 9.1, a_0 is calculated in the last column from Eq. (9.21) as $a_0 = v_r^2 / R_{rim}$. The mean value of a_0 for the spiral galaxies in Table 9.1 is 0.061 ± 0.011 nm/s^2.

For a star at a radius R_{rim} corresponding to the visible rim of one of these galaxies, $a_0 \approx 0.06$ nm/s^2 is about $0.1D$, where $D = 0.69 \pm 0.06$ nm/s^2 is the universal inertial drag constant from Eq. (6.13). This comparison of a_0 and D

68

Table 9.1. Velocity v_r, radius R_{rim}, and acceleration constant $a_0 = v_r^2 / R_{\text{rim}}$ at the visible rim of each galaxy listed. Rotation curve data adapted from [32].

Galaxy	v_r (km/s)	R_{rim} (kpc)	a_0 (nm/s^2)
U 2259	80	4.0	0.052
N 2403	130	9.5	0.058
N 3198	150	11	0.065
N 5907	220	20	0.078
N 4565	230	25	0.068
U 2885	300	63	0.045

shows that inertial drag dominates gravity and centripetal acceleration in the outer regions of all pressure-supported isothermal spheres, and including spiral galaxies, a topic for Ch. 10.

Many spiral galaxies show evidence of flat rotation curves in the outer regions beyond R_{rim}, the radius at the rim of a pressure-supported isothermal sphere beyond which a significant number of stars in random thermal motion escape to infinity. The evidence appears in the 21-cm lines of hydrogen gas clouds in these outer regions as either a dispersion velocity or rotation velocity. Flat rotation curves beyond R_{rim} are not accounted for by a model of a spiral galaxy as a pressure-supported isothermal sphere. However, to the extent that spiral galaxies are not spherically symmetric mass distributions, rotation curves should not be expected to fall as $v(r) \sim r^{-1/2}$ beyond R_{rim}.

In the regions beyond R_{rim}, the density might be expected to decay exponentially with a scale length much shorter than R_{rim}. Appendix C calculates the enhancement of the gravitational field in the *plane of a disk galaxy* when the galaxy is modeled as a thin disk, rather than a spherically symmetric mass. Figure C.5 in Appendix C shows the rotation curves calculated for several truncated exponential disks of different scale heights. This figure shows that some such exponential disks have flat rotation curves according to unmodified Newtonian dynamics. This result suggests that the disk-like geometry may be a significant factor in extending the region of flat rotation curves beyond the visible rim of pressure-supported spiral galaxies.

Chapter 10
Gravitational Reach and Inertial Drag in Cluster, Galaxy, and Gas Dynamics

This section calculates the effects of inertial drag on the gravitational encounter of small particles with a much larger mass, such as the encounter of gas particles with the core of a galaxy. For the purpose of this calculation, the much larger mass is treated as a point mass m_0. The calculation treats the encounter between the mass and particles as a nonrelativistic, weak-field interaction. That is, the interaction involves only Newtonian dynamics, except that the effects of inertial drag are included in the dynamics.

This section also calculates the deceleration by inertial drag of the center of mass of a pressure-supported system as a function of its drift speed V and temperature T as it drifts through the static rest frame F of an inertial field. Hotter systems are decelerated less by inertial drag than colder ones. The model has applicability to systems of particles such as molecules in galactic gas clouds and galactic coronas, stars in globular clusters and galaxies, and galaxies in galactic clusters.

Larger and hotter galaxies and other pressure-supported systems may be expected to decelerate more slowly with respect to the rest frame F than their smaller and cooler counterparts. In a hot pressure-supported system, more of the distance travelled through F by particles is due to internal motion than in colder systems. More of the constant inertial drag force, therefore, acts to 'cool' the internal motion of a hot system, leaving less drag to impede the motion of the center of mass.

Deceleration of pressure-supported systems by inertial drag is estimated according to the model of isothermal spheres of Ch. 9 as follows. Sources and sinks of energy, such as heating, radiation, compression, and exchange of mass with a surrounding medium, are ignored in this model. For calculating the instantaneous deceleration of the center of mass of an isothermal sphere by this model, the gravitational field within the system is neglected, as is rotation. The system is characterized by just two parameters, an isothermal temperature T (in energy units) and a drift velocity of its center of mass through the rest frame F

of the universe of $\mathbf{V} = V(t)\hat{\mathbf{x}}$ in the x direction, affected only by inertial drag.

In this simple model, the total specific energy of the system in the rest frame F is

$$E = \frac{1}{2}\sum_{i=1}^{N}\left[(\dot{x}_i + V)^2 + \dot{y}_i^2 + \dot{z}_i^2\right] = E_0 - D\sum_{i=1}^{N} s_i, \tag{10.1}$$

where $\mathbf{r}_i = x_i\hat{\mathbf{x}} + y_i\hat{\mathbf{y}} + z_i\hat{\mathbf{z}}$ is the displacement vector of the ith particle from the center of mass, E_0 is the initial specific energy of the system at time $t = 0$, and

$$s_i = \int_0^t \left[(\dot{x}_i + V)^2 + \dot{y}_i^2 + \dot{z}_i^2\right]^{1/2} dt \tag{10.2}$$

is the total distance travelled by the ith particle in the rest frame F beginning at $t = 0$.

Assuming a Maxwellian distribution of velocities in the center-of-mass frame, the total specific energy of the system averaged over all N particles is $N(V^2 + v_{\text{rms}}^2)/2$, where v_{rms} is the root-mean-square (rms) speed of the particles in the center-of-mass frame. For an ideal gas, $v_{\text{rms}}^2 = 3T/m$, and v_{rms}^2 and $3T/m$ may be used interchangeably to refer to the 'thermal' specific energy per particle, that is, the specific kinetic energy per particle in the center-of-mass frame, whether the particles are molecules in a gas or galaxies in a cluster.

Within the approximations of this model, the instantaneous deceleration of the center of mass of the system in the rest frame F as a function of its speed in F is

$$d\ln(V)/dt \approx -D/(V^2 + v_{\text{rms}}^2)^{1/2} \approx -D/(V^2 + 3T/m)^{1/2}, \tag{10.3}$$

and is shown in Fig. 10.1.

Over a time short compared to the time over which the temperature or speed or gravitational potential of the pressure-supported system changes appreciably, the velocity of the center of mass of the system in the rest frame F then becomes

$$\mathbf{V} \approx V_0\left[1 - Dt/(V_0^2 + v_{\text{rms}}^2)^{1/2}\right]\hat{\mathbf{x}}, \tag{10.4}$$

where V_0 is the speed at $t = 0$. Although many factors enter into a calculation

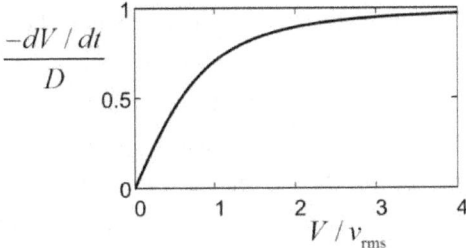

Fig. 10.1. Normalized instantaneous deceleration of isothermal sphere *vs.* instantaneous speed of center of mass in F, normalized to rms velocity of particles.

of the effective temperature of a pressure-supported system, a rough estimate of the instantaneous cooling rate of the system due solely to inertial drag is

$$d \ln(T) / dt \approx -2D / (V^2 + 3T/m)^{1/2}. \tag{10.5}$$

Equation 10.3, the instantaneous deceleration of a pressure-supported isothermal sphere in an inertial field, has some application to an analysis of the Bullet cluster. The term 'Bullet cluster' commonly refers to a system comprising four components – two galaxy clusters and two hot, x-ray-emitting plasma clouds. The image of the plasma clouds was observed with the Chandra X-ray observatory in 2002–2004. A contour map of gravitational potentials was derived by [45, 46] from weak-gravitational-lensing data [45].

The usual interpretation given to the Bullet cluster is that it is the aftermath of a collision of two galaxy clusters, a main cluster and a smaller sub-cluster, that merged and then separated. Because the galaxies within the clusters are effectively collisionless particles, the main cluster and sub-cluster passed through each other with little effect on the motion of their centers of mass. In this interpretation, the two plasma clouds are collisional, and were slowed by ram pressure, thereby lagging behind the clusters in which each had been embedded. Furthermore, in this interpretation, the shape of the plasma cloud associated with the sub-cluster is that of a Mach 3 bow shock [47], from which a separation speed of the sub-cluster from the main cluster of 4700 km/s is inferred [45, 46].

In this interpretation, the significance of the Bullet cluster is that it seems to offer "direct empirical proof of the existence of dark matter [45]." Dark matter particles, generally assumed to be collisionless, would be expected to remain gravitationally bound to their associated clusters of galaxies and to pass through the merger, as would the galaxies, without slowing. The peaks in the contours of gravitational potentials, which correspond to peaks in surface mass density, are located at the presumed cluster centers, rather than in the plasma clouds, even though the plasmas in galaxy clusters are generally much more massive than visible galaxies in clusters. In this interpretation, the peaks in mass density outside the plasma clouds suggests that much of the mass within the galaxy clusters is not visible, and therefore represents dark matter. Regardless of which interpretation of the Bullet cluster ultimately prevails, the achievements and observational and theoretical advances underlying the analysis by [46] are notable.

Some interpretations suggest that ordinary baryonic matter with modified dynamics like MOND might account for the missing mass. For example, [48] notes that the missing mass for which MOND does not account in the two clusters of the Bullet cluster system is a discrepancy of only about a factor of two, and proposes that ultra-diffuse galaxies within the clusters could account for that missing mass.

It could be that neither MOND nor dark matter is needed to interpret the Bullet cluster. For example, the four components of the Bullet cluster or some combination of them might be independent and unrelated, and might just coincidentally appear to be close to each other along the line of sight. The main cluster and sub-cluster are at a distance greater than 1 Gpc, and appear to be separated by a distance of 720 kpc projected onto a plane normal to the line of sight. The actual separation might be much greater than 720 kpc.

From the speed of cluster separation of 4700 km/s deduced from the 'bow shock' of the smaller plasma cloud, the cores of the clusters are supposed to have passed through each other more than 100 million years ago [45]. This interpretation raises some questions. Typical dispersion velocities of galaxies within clusters are less than or about 1000 km/s, and typical peculiar velocities of astrophysical objects throughout the universe are only about a few hundred km/s with respect to their backgrounds, an order of magnitude slower than the supposed separation speed within the Bullet cluster. How did the largest gravitationally bound object in the universe, a cluster of galaxies, come to acquire a peculiar velocity an order of magnitude greater than typical peculiar velocities throughout the universe? This question is addressed within the context of ΛCDM and MOND by [49].

If the bow-shock-shaped plasma cloud passed through the larger plasma cloud, and both plasmas were collisional enough to slow down their separation speed, why is there no evidence, other than possibly the bow-shock shape, of such a colossal event as the transit of one through the other? If each plasma cloud before a merger was gravitationally bound and collisional, should not one expect a much greater gravitational interaction between the clouds during their merger than seems to be in evidence after the merger? And if the clusters are filled with a dark mass much greater than the mass in the plasma clouds, how were the plasmas able to become gravitationally unbound from their respective clusters? By the same token, how was the visible mass in the clusters able to become gravitationally unbound from the much more massive plasma clouds?

If the possibility of inertial drag is considered, questions over the interpretation of the Bullet cluster multiply. According to Eq. 10.3, a cluster separation speed of $V_s = 4700\,\text{km/s}$ should be decelerated over a time of order $V_s / D \approx 200\,\text{Myr}$ to a much slower separation speed, and during that time the clusters should reach a separation of order $V_s^2 / 2D \approx 510\,\text{kpc}$. That is, even if the cores of the clusters started their separation from the same location at a speed of 4700 km/s, they would not have reached a separation of 720 kpc. According to the generally accepted interpretation of the Bullet cluster, however, the clusters began by approaching each other before their merger. Allowing for inertial drag, then, the initial approach speed must have been substantially faster than 4700 km/s in this interpretation.

An alternative way in which the plasma cloud of the Bullet cluster could have acquired its bow-shock shape is illustrated by the sketch in Fig. 10.2. When the cloud began moving through F, if it had a hot central core and a falling radial temperature profile, as indicated schematically in Fig. 10.2 with temperatures $T_1 > T_2 > T_3$, then according to Eq. (10.3), the hot central core should have been decelerated less by inertial drag than the cooler outer regions of the cloud.

The remainder of this section treats the effects of inertial drag on the interaction between a large point mass and a cloud of particles. The effects differ depending on the relative velocities of the mass and the particle cloud with respect to the rest frame F of a static universe. As was done in Ch. 8 for orbital

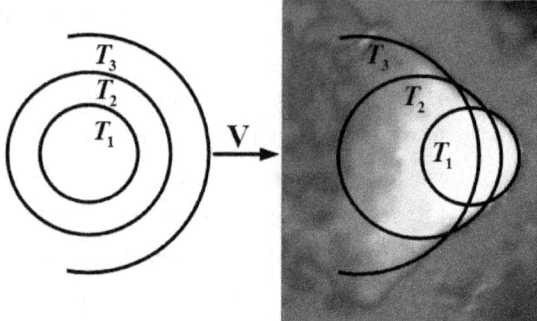

Fig. 10.2. Inertial drag causes less deceleration of hotter plasma, which could account for the bow-shock shape in the Bullet cluster.

dynamics, this section presents calculations for two limiting cases: (1) A cloud of particles moving with speed V_0 through the rest frame F of a static universe encounters a large point mass m_0 at rest in F; and (2) a point mass moving with speed V_0 through the rest frame F of a static universe encounters a particle cloud at rest in F. Before either calculation is presented, however, the concept of the 'gravitational reach' of a mass is introduced.

The gravitational reach of a mass m_0 in F is defined as the range from m_0 beyond which a particle *at rest* is not caused to move by the gravitational field of m_0. The situation is similar to that of a mass resting on a surface with friction: The mass will not move until the force on it exceeds a threshold. Chapter 6 showed that every particle at rest in the rest frame F of a static universe is 'locked' in position by the inertial field at the bottom of its own potential well. Equation (6.12) showed that for small virtual displacements of a particle at rest in F, the specific inertial drag force opposing any displacement of the particle at rest is $-D$. That is, unless the gravitational force of m_0, or any force, on a particle at rest in F exceeds the inertial drag force, the particle must remain at rest, 'locked' in position. The gravitational reach of a mass m_0, therefore, must be the range at which its Newtonian gravitational specific force on a particle at rest equals the constant inertial drag specific force, $-D$, opposing displacement of the particle. In terms of the mass of our Sun, $M_\odot = 1.99 \times 10^{30}$ kg, the gravitational reach of a mass m_0 is

$$R_g \equiv (Gm_0 / D)^{1/2} = (4.40 \pm 0.19) \times (10^{14} \text{ m})(m_0 / M_\odot)^{1/2}$$
$$= (1.40 \pm 0.06) \times (10^{-2} \text{ pc})(m_0 / M_\odot)^{1/2} \tag{10.6}$$

where the inertial drag constant was taken to be $D = 0.69 \pm 0.06 \text{ nm} / \text{s}^2$ from Eq. (6.13).

Even the largest galaxies, with a mass of $10^{12} M_\odot$, have a gravitational reach only of order 10 kpc. The Milky Way, with a mass of a few times $10^{10} M_\odot$, is thought to have a radius of 15 to 20 kpc. That is, the radius of a typical spiral galaxy, like our Milky Way, is greater than its gravitational reach, and inertial drag can be expected to contribute significantly with the Newtonian gravitational field to the rotational dynamics at the outer edge of the galaxy.

Related to the concept of gravitational reach is the 'escape velocity', $v_e(r)$, at a range r from a mass m_0. From Eq. (B.11), a nonrelativistic particle launched radially outward from a spherical mass m_0 at rest in F at the origin satisfies the equation of motion,

$$\ddot{r} = -Gm_0 / r^2 - D. \tag{10.7}$$

(If the particle is moving radially *inward*, the equation becomes $\ddot{r} = -Gm_0 / r^2 + D$.) The escape velocity is defined as the radial velocity that must be imparted to the particle at $r < R_g$ in order for the particle to just reach R_g, where it remains motionless, locked in position. From Eq. (10.7), the escape velocity, shown in Fig. 10.3, is

$$v_e(r) = v_{e0}(r)[1 - r^2 / R_g^2]^{1/2} \text{ for } r \le R_g, \tag{10.8}$$

where, in the absence of inertial drag, $v_{e0}(r) \equiv (2Gm_0 / r)^{1/2}$ is the escape velocity from r to infinity and $v_{e0}(R_g) \equiv (4Gm_0 D)^{1/4}$ is the escape velocity from R_g to infinity. On the other hand, with inertial drag, if the particle is already at rest beyond the gravitational reach of m_0, that is, for $r > R_g$, then the particle has already 'escaped', and the escape velocity in that case is zero.

If a particle is moving beyond the gravitational reach of m_0, or if m_0 is moving in the rest frame F, then the gravitational field of m_0 may still influence the motion of the particle. From Eq. (B.11), a nonrelativistic particle moving with velocity $\mathbf{v}$ with respect to a spherically symmetric mass m_0 at the origin, and assumed to be moving entirely outside the mass m_0, satisfies the equation of

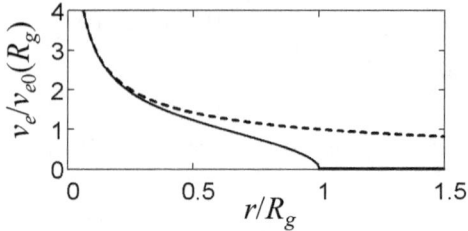

Fig. 10.3. Normalized escape velocity vs. radius normalized to R_g for particle starting from radius r with inertial drag (solid) and without (dashed).

motion,

$$dv / dt + (Gm_0 / r^2)\hat{\mathbf{r}} = -D\hat{\mathbf{s}}, \tag{10.9}$$

where $\hat{\mathbf{s}} \equiv (\mathbf{v} + \mathbf{V_0}) / |\mathbf{v} + \mathbf{V_0}|$ is a unit vector in the direction of particle motion through the rest frame F, and $\mathbf{V_0}$ is the constant velocity of m_0 through F. In this section, as in Ch. 8, there is no difference between the specific momentum of the particle, $\dot{\mathbf{s}} = d\mathbf{s} / d\tau$, and its coordinate velocity, $d\mathbf{s} / dt$.

In Ch. 8, Eq. (10.9) was treated for the two limiting cases, $\mathbf{V_0} = 0$ and $\mathbf{V_0} \gg \mathbf{v}$. For treating inertial drag in galactic dynamics, any of the acceleration terms, dv / dt, Gm_0 / r^2, D, v^2 / r, and V_0^2 / r, may be comparable to or greater than others. This section will treat inertial drag beyond the mass distribution, so that m_0 is effectively a point mass.

Generally speaking, as seen from Ch. 8 and from Eq. (10.3), if a galaxy is moving through the rest frame F of a static universe with a speed V_0, the deceleration of particle motion by inertial drag within the galaxy is reduced by a factor of order $v / (v^2 + V_0^2)^{1/2}$, and the effective gravitational reach of the galaxy is extended by a factor of the order of $[(v^2 + V_0^2)^{1/2} / v]^{1/2}$. Case 1 presents the calculation of a particle moving with speed V_0 through the rest frame F of a static universe and encountering a point mass m_0 at rest in F.

Case 1. A Moving Particle Encounters a Point Mass at Rest in F

As in the first case treated in Ch. 8, the model for this calculation features a central point mass m_0 at rest in F, and features nonrelativistic, weak-field Newtonian dynamics with inertial drag. A central point mass might represent a large compact body about which a galaxy forms from a streaming interstellar gas.

In terms of the specific angular momentum of the orbiting particle, $l = r^2 \dot{\phi}$, the radial and azimuthal components of Eq. (10.9) describing the motion of a particle about a central point mass m_0 become

$$\ddot{r} - l^2 / r^3 + Gm_0 / r^2 = -D\dot{r} / v, \tag{10.10}$$

$$\dot{l} = -Dl / v, \tag{10.11}$$

where the coordinate speed is $v = ds / dt = (\dot{r}^2 + l^2 / r^2)^{1/2}$. The specific energy of the particle is

$$E = -Gm_0 / r + v^2 / 2 = E_0 - Ds, \tag{10.12}$$

where E_0 is the initial specific energy of the particle at $s = 0$. If the inertial drag constant D is comparable to or greater than Gm_0 / r^2 or v^2 / r, then inertial drag will have a substantial effect on the orbital dynamics.

To gain insight into the behavior of a moving particle encountering a galaxy at rest in F, consider a particle that initially has azimuthal speed v_0 outside the galaxy at distance r_0 from the galaxy center and has no radial velocity, so the specific angular momentum of the particle initially is $l_0 = r_0 v_0$. Two dimensionless constants are defined with respect to the initial motion,

$$A_0 \equiv (v_0^2 / r_0) / D, \tag{10.13}$$

$$B_0 \equiv (v_0^2 / r_0) / (Gm_0 / r_0^2). \tag{10.14}$$

Here, A_0 is the ratio of initial specific centripetal force to specific inertial drag force acting on the particle, and B_0 is the ratio of initial specific centripetal force to initial specific gravitational force. Since A_0 is related to B_0 by $B_0 / A_0 = (r_0 / R_g)^2$, the ratio B_0 / A_0 is a measure of the relative strength of inertial drag to gravity at r_0.

An analysis of the equations of motion, Eqs. (10.10) and (10.11), shows that a particle with these initial conditions will undergo one of two types of motion, depending on the values of A_0 and B_0. Either the particle will be captured into a bound orbit of ever decreasing radius, or it will not be captured. If it is not captured, either the particle will escape to infinity or it will come to rest outside

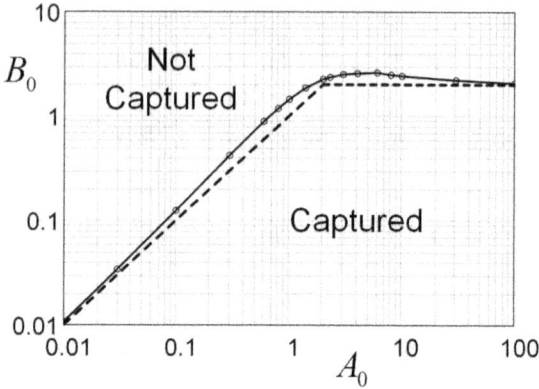

Fig. 10.4. Particles with initial values (A_0, B_0) below the separatrix (solid curve) are captured by a central mass at rest in F; those above are not. Dashed lines are asymptotes of the separatrix.

the gravitational reach of the mass m_0. The solid curve in Fig. 10.4 is the separatrix of the two modes of behavior of the particle, captured or not captured.

The dashed lines in Fig. 10.4 represent the asymptotic values of the separatrix in the limits $A_0 \to \infty$ and $A_0 \to 0$. If v_0 is so fast that inertial drag is negligible compared to the centripetal force on the time scale of one orbit, then $A_0 \gg 1$, and from the energy equation, Eq. (10.12), the particle orbit will be bounded if $B_0 < 2$ and unbounded if $B_0 > 2$. If v_0 is so slow that inertial drag is overwhelming on the time scale of one orbit, then $A_0 \ll 1$, and from the radial equation of motion, Eq. (10.10), the particle will fall inwards if the particle is within the gravitational reach of the galaxy, that is, if $B_0 < A_0$, and will come to rest outside the gravitational reach of the galaxy if $B_0 > A_0$.

Figure 10.5 shows the difference in behavior of particle motion for slow particles ($A_0 \ll 1$) and fast particles ($A_0 \gg 1$) that have values of B_0 on the separatrix between being captured and not captured. That is, the particles come to rest exactly at the gravitational reach of the galaxy that is at rest in F. The slow particle in Fig. 10.5(a) falls inward until it is stopped by inertial drag at the gravitational reach. The fast particle in Fig. 10.5(b) first moves outward and then falls inward until it is stopped by inertial drag at the gravitational reach.

Figure 10.6 shows trajectories of particles outside a galaxy at rest in F for four different values of B_0. Here the particles are fast enough initially that inertial drag is much less than centripetal acceleration ($A_0 = 10$), but not negligible.

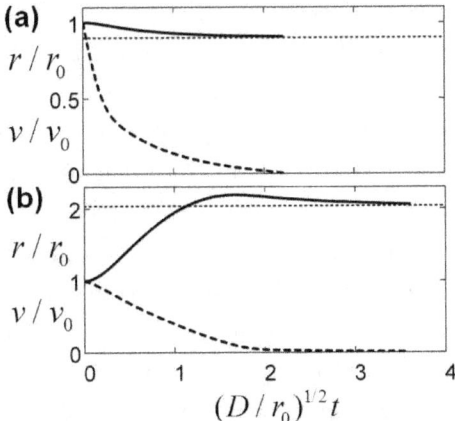

Fig. 10.5. Normalized distance (solid curve) from center of galaxy at rest in F and normalized speed (dashed) vs. normalized time of particle on the separatrix between being captured and not captured for: (a) $A_0 = 0.1$; and (b) $A_0 = 10$. Dotted line is normalized gravitational reach R_g / r_0.

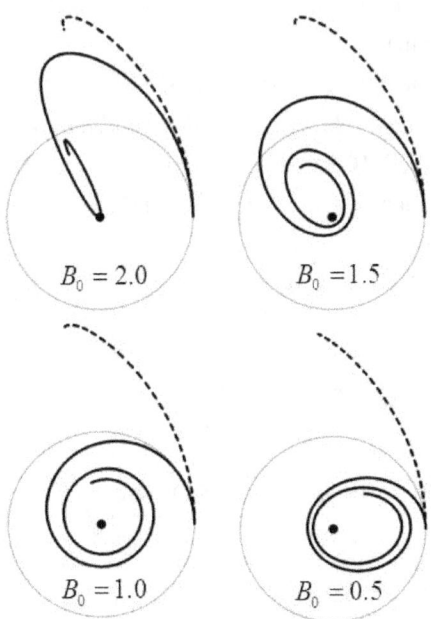

Fig. 10.6. Particle trajectories (solid curves) about central mass (dot) for $A_0 = 10$ and values of B_0 indicated. Dashed curve is trajectory of particle on separatrix ($B_0 = 2.42$) between captured and not captured.

Case 2. A Point Mass Encounters a Cloud of Particles at Rest in F

In this case a point mass m_0 moving with speed V_0 through the rest frame F of a static universe encounters a cloud of particles at rest in F. If a density gradient normal to the trajectory of the point mass in F exists in the particle cloud, then particles will be accreted onto the point mass preferentially with one sign of angular momentum. As in Case 1, the model features nonrelativistic, weak-field Newtonian dynamics with inertial drag. In this case, only the gas that is within the gravitational reach of the point mass along its trajectory through F will be affected. The point mass and the particles, once they start moving, are slowed by inertial drag. The mass of the particles is taken to be negligible compared to the point mass, so that the point mass is not slowed by accretion of particles or by energy loss to the particles.

Trajectories of particles within the gravitational reach of the point mass m_0 are calculated from the model shown in Fig. 10.7. From this configuration, trajectories are calculated for particles initially at rest at varying impact parameters y_0 from the x-axis, along which the point mass m_0 is moving through F with initial velocity $\mathbf{V}_0$. Particles do not begin moving until they are within the gravitational reach of the mass m_0.

In the rest frame F, the position of the point mass on the x-axis is $X(t) = V_0 t - Dt^2 / 2$ for $t \le V_0 / D$, at which time the point mass is stopped by inertial drag. From Eq. (10.9), the equation of motion for each particle, once it is within the gravitational reach of m_0, has components

$$\ddot{x} = \frac{-Gm_0[x - X(t)]}{\left([x - X(t)]^2 + y^2\right)^{3/2}} - 2D\frac{\dot{x}}{v} \quad,$$

$$\ddot{y} = \frac{-Gm_0 y}{\left([x - X(t)]^2 + y^2\right)^{3/2}} - 2D\frac{\dot{y}}{v} \quad,$$

$$(10.15)$$

where $v \equiv (\dot{x}^2 + \dot{y}^2)^{1/2}$ is the speed of the particle in F. In terms of the normalized variables,

$$\tilde{x} \equiv x / R_g \, , \; \tilde{y} \equiv y / R_g \, , \; \tilde{t} \equiv V_0 t / R_g \, , \; \tilde{v} \equiv v / V_0 \, ,$$

$$\tilde{r} \equiv \left[(\tilde{x} - \tilde{t} + K_{TF} \tilde{t}^2 / 2)^2 + \tilde{y}^2 \right]^{1/2} \, ,$$

$$(10.16)$$

the components of the equation of motion become

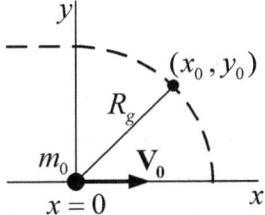

Fig. 10.7. Configuration at $t = 0$ for calculating trajectories of particles initially at rest at (x_0, y_0), in field of point mass m_0 moving through F with initial velocity $\mathbf{V_0}$.

$$\frac{d^2\tilde{x}}{d\tilde{t}^2} = -K_{TF}\left[\frac{\tilde{x} - \tilde{t} + K_{TF}\tilde{t}^2/2}{\tilde{r}^3} + \frac{d\tilde{x}/d\tilde{t}}{\tilde{v}}\right].$$

$$d^2\tilde{y}/d\tilde{t}^2 = -K_{TF}\left[\tilde{y}/\tilde{r}^3 + (d\tilde{y}/d\tilde{t})/\tilde{v}\right]$$

(10.17)

In the normalized variables of Eq. (10.17), the initial conditions for each particle are taken to be $\tilde{y}_0 = y_0/R_g$, $\tilde{x}_0 = (1 - \tilde{y}_0^2)^{1/2}$, and $\tilde{v}_0 = 0$ at $\tilde{t} = 0$. Figure 10.7 showed the initial configuration for Case 2, a point mass m_0 encountering a particle at rest in F at an impact parameter y_0.

In these normalized variables, the equation of motion, Eq. (10.17), is parametrized solely by the dimensionless constant,

$$K_{TF} \equiv (DGm_0/V_0^4)^{1/2} .$$

(10.18)

The value of this 'Tully-Fisher constant' K_{TF} determines whether a particle at rest encountered by the point mass m_0 is captured by the mass or not. For any initial impact parameter $y_0 < R_g$, there is a critical value of the constant K_{TF}, below which a particle initially at rest in F will not be captured into a bound orbit of m_0. For K_{TF} below this critical value, the point mass m_0 is moving so fast in F that by the time m_0 is stopped by inertial drag, after moving a distance $R_g/2K_{TF}$ along the x axis, the particle is stopped by inertial drag outside the gravitational reach of m_0.

Just as in Case 1, a particle will undergo one of two types of motion, depending in this case on the values of the initial impact parameter y_0 and K_{TF}. Either the particle will be captured into a bound orbit of ever decreasing radius about m_0, or it will come to rest outside the gravitational reach of m_0. The solid curve in Fig. 10.8 is the separatrix of the two modes of behavior of the particle, captured or not captured.

As the impact parameter of the particles, y_0, increases above about $0.4R_g$, the magnitude of K_{TF} must increase in order to capture the more distant particles. That is, the point mass must be heavier or slower. But as seen in Fig. 10.8, as y_0 decreases below about $0.4R_g$, the magnitude of K_{TF} must also increase in order to capture the more distant particles. The reason is that the point mass is so heavy and slow that the slingshot effect propels the particles beyond the gravitational reach of the mass. An example of the slingshot effect on the orbits is seen in Fig. 10.9 for an impact parameter, $0.4R_g$.

Figure 10.9 shows examples of orbits of particles that are captured and not captured. Both particle orbits in this figure began at the same impact parameter, $y_0 = 0.3R_g$. The point mass had $K_{TF} = 0.18$, which was just above the separatrix, for the captured particle, and $K_{TF} = 0.17$, just below the separatrix, for the particle that is not captured. The slingshot effect caused the latter particle to be propelled beyond the gravitational reach of the point mass after it had been stopped by inertial drag.

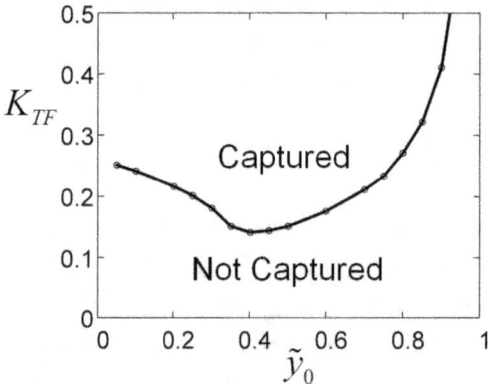

Fig. 10.8. Particles with initial normalized impact parameter $\tilde{y}_0 < 1$ in the field of a point mass with K_{TF} above the separatrix (solid curve) are captured by the mass; those below are not.

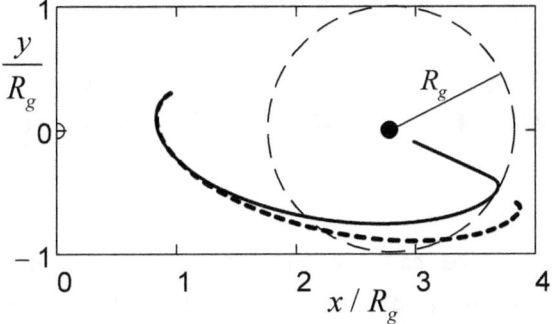

Fig. 10.9. Orbits of particles initially at rest in F with impact parameter $y_0 = 0.3R_g$ in field of point mass having $K_{TF} = 0.18$ (solid curve) and $K_{TF} = 0.17$ (dashed curve).

Chapter 11
Dipole Inertial Radiation from Unbound Quadrupoles

This section summarizes earlier calculations of dipole radiation emitted from gravitationally unbound quadrupoles [9]. These calculations are used in Ch. 12 to show that the momentum and power radiated by an accelerating mass are consistent with the reaction force of Newton's third law, "To every action there is always opposed an equal reaction."

Dipole gravitational disturbances from gravitationally unbound mass quadrupoles propagate to the radiation zone with signal strength at least of quadrupole order if the mass components of the quadrupoles are nonrelativistic, and of dipole order if relativistic [9]. Gravitationally unbound mass quadrupoles, despite having constant mass dipole moment, nevertheless produce dipole perturbations of gravitational fields that do not completely destructively interfere as they propagate to the radiation zone, that is, to distances much greater than the perturbation wavelengths. Even in slow-speed, weak-field systems, first-order relativistic effects, such as phase differences and frequency shifts, prevent complete destructive interference of the dipole perturbation fields of unbound quadrupoles, and allow dipole inertial waves with signal strength of quadrupole order to reach the radiation zone. The signal strength of dipole inertial waves from relativistic-speed quadrupoles will generally be of dipole order, and many orders of magnitude greater, owing to a complete disruption of the interference of the dipole perturbation fields.

In [9], the interfering dipole fields from unbound quadrupoles that propagate to the radiation zone were named dipole *gravity* waves and the power transported to the radiation zone was said to be carried by dipole *gravitational* radiation. In this section the interfering dipole fields are named dipole *inertial* waves and the power transported to the radiation zone is said to be carried by dipole *inertial* radiation. Both nomenclatures are technically correct. Chapter 3 of this book showed that the gravitational field of a local mass is just the contribution of that mass to the 'sum for inertia' and to the inertial field of the universe. Since this section and the next are concerned primarily with fields in the radiation zone of accelerating masses, where contributions to the inertial field are significant and where the fields fall with range as r^{-1}, the latter nomenclature, dipole *inertial* fields, is used here.

A dipole inertial wave is just a gravitational disturbance propagating at light speed that can cause the mass dipole moment of a detector to change. Dipole inertial waves can cause an isolated particle to move by pushing or pulling it in the direction of polarization. In contrast, classical (transverse traceless) quadrupole gravity waves do not cause an isolated particle to move.

In this section, amplitudes of dipole inertial waves are calculated using a weak-field approximation of general relativity. In this approximation, terms of order Φ^2 are neglected, where Φ is a characteristic gravitational potential in the quadrupole. This weak-field approximation cannot be used, therefore, to calculate general relativistic effects for gravitationally bound quadrupoles, such as binary star systems. Such low-order general relativistic effects as periastron precession depend on potential terms of order $\beta^2\Phi$, where β is a characteristic speed divided by the speed of light c. In a gravitationally bound rotating quadrupole, $\beta^2\Phi$ is of the same order as Φ^2/c^2. The calculations in this section, therefore, are valid only for gravitationally unbound quadrupoles satisfying $\Phi/c^2 \ll \beta^2$.

Another way to understand this weak-field approximation is that field terms of order G^2 are neglected. That is, the field strength is Newtonian even though the field is relativistically exact to all orders of β. In fact, the field reduces to the Newtonian field in the limit $\beta \to 0$.

Any of the following conditions can prevent complete destructive interference of dipole inertial fields in the radiation zone from a source having constant dipole moment: (i) A difference in ranges to the detector of the quadrupole elements, causing the observed phase difference to be shifted from π; (ii) a difference in velocities of the quadrupole elements, causing the observed waves to be frequency-shifted differently; and (iii) a source modification that acts on the fields of the individual quadrupole elements differently within the source region.

There has been some question as to what happens to the dipole power emitted by each dipole in a system with zero net dipole moment, or even whether the dipoles under such conditions emit at all. A fourth-order calculation (to order β^4) of the retarded electric field in the source region of a *nonrelativistic periodic* system of charges showed that all accelerated charges emit dipole radiation at the Larmor rate, but that if the system has zero net dipole moment, the emitted dipole power is reabsorbed by doing work on the other charges in the system [50]. The implication of the conclusions from [50] and [51] is that every accelerated mass produces a local dipole perturbation of the gravitational field, which is just what is expected from the most elementary considerations.

The following example in this section illustrates how all three causes of incomplete interference mentioned above can contribute to dipole inertial perturbations reaching the radiation zone. Consider an oscillator, such as a pendulum or mass on a spring, fixed to the Earth's surface. The net dipole moment of the oscillator-plus-Earth system is zero. The mass dipole moment of the oscillator is balanced not by another rigid mass, however, but by stress waves propagating through the Earth. The dipole gravitational perturbation of the oscillator has a signature very different from that of the stress waves and can be measured independently, at least in the source region. The sample calculation in this section will show that dipole perturbations produced by this system do not destructively interfere completely, even in the radiation zone.

The one-dimensional system illustrated in Fig. 11.1 comprises a mass m connected by a lossless light spring to a much more massive long rod that extends in the z direction and has negligible motion and negligible energy attenuation. The oscillation amplitude of the mass is a, and the angular velocity is ω. The position of the mass is $z = a \cos \omega t$, and its dipole moment per unit length is $mz\delta(z - a\cos\omega t)$, where δ is the delta function. Here, $\hat{n}_{\parallel} \equiv \hat{n}$ is the unit vector in the direction parallel to propagation of the wave towards a detector at $\mathbf{r}_0$, and $\hat{n}_{\perp} \equiv \hat{n} \times (\hat{n} \times \hat{z}) / \sin\theta$ is a unit vector in the transverse direction.

The oscillation of the mass is balanced primarily by an elastic compression wave, or P-wave. The compression wave propagates up the rod with speed ω / k, where k is its wave-number. The dipole moment per unit length of the compression wave is $-mak\cos(kz)\delta(kz - \omega t)$. The combined momentum per unit length of the oscillating mass and the compression wave,

$$\dot{D} = ma\omega\left[z\sin(\omega t)\tilde{\delta}(z - a\cos\omega t) + k\cos(kz)\tilde{\delta}(kz - \omega t) \right] \quad , \qquad (11.1)$$

is the time derivative of the combined dipole moment per unit length, where $\tilde{\delta}$ is the derivative of the delta function with respect to its argument.

The combined momentum of the system is the integral of the momentum density along the z axis, $\int \dot{D} dz$, which is a constant. The combined dipole mo-

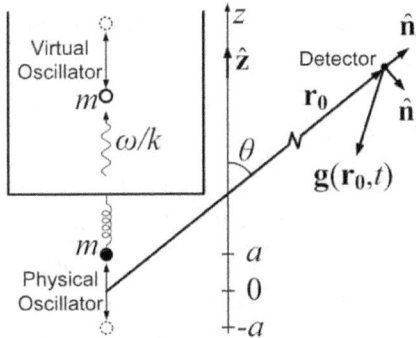

Fig. 11.1. Model configuration of gravitationally unbound quadrupole with zero dipole moment that radiates dipole inertial waves to radiation zone at $\mathbf{r}_0$.

ment, which is the time integral of the combined momentum, has zero second derivative with respect to time, that is, $\ddot{D} = 0$. Although the momentum of the system is constant, the system has a time-dependent and non-periodic distribution of momentum density along the z axis, owing to the modification of the source region by the elastic rod. Both components of the momentum density oscillate periodically, one about $z = 0$, the other about $z = \omega t / k$. This non-periodic time dependence of the momentum density further disrupts the interference of the dipole inertial perturbations of the system, as seen from almost all distant observation points. The dipole inertial power radiated from this system is calculated in this section.

For the model in Fig. 11.1, no additional energy or momentum is required to propagate the compression wave in the rod once the oscillator has been set in motion initially, assuming the rod is massive enough that it does not move appreciably and long enough that the wave is not reflected. For half of each wave cycle, the oscillator does work on the compression wave in the rod. For the other half of each wave cycle, the compression wave in the rod does equal work on the oscillator. A nearly lossless oscillator at the end of a long, massive rod will oscillate with nearly constant amplitude, independently of the extent of the compression wave it generates in the rod. The compression wave does not transport energy or momentum, except in the first fraction of a wave cycle associated with the oscillator start-up, as demonstrated by Eq. (11.1), which shows that the momentum density of the compression wave is concentrated entirely at the front of the wave at $z = \omega t / k$. That is why the integral of the momentum density along the z axis is a constant and why the momentum and energy of the compression wave can be emulated by a single virtual oscillator moving at the front of the physical wave.

Figure 11.2 shows a world-line representation of the model in Fig. 11.1 of an oscillator/stress-wave pair. The center of mass of the system, represented by the straight world line 'CM', is unaccelerated, but the constituent mass dipoles of the system, represented by the oscillating world lines, are accelerated. The physical oscillator oscillates about the origin. The virtual oscillator oscillates about the point $z = (\omega / k)t$ at the front of the stress wave. Originally, the dipole fields of the physical and virtual oscillators are out of phase and destructively interfere everywhere.

The inertial field in the radiation zone is a linear superposition of the retarded field amplitudes and phases of each constituent of the quadrupole. The superposition field depends on the spacetime point at which it is observed. As shown in Fig. 11.2, the superposition field at a fixed detector of two unbound dipoles will generally not completely destructively interfere and will not vanish at all retarded times. In fact, Fig. 11.2 shows the sources at a moment when their retarded fields are in phase and *constructively* interfere at a detector.

Within the weak-field approximation of general relativity, [4] derived an exact expression for the retarded gravitational field of a mass in arbitrary relativistic motion,

$$\mathbf{g}(\mathbf{r},t) = -Gm\left\{ \frac{\alpha\hat{\mathbf{n}} + [(2\gamma^2 + 1)\kappa - 4]\gamma\boldsymbol{\beta}}{\gamma^2\kappa^3 R^2} \right.$$
$$\left. + \frac{(\hat{\mathbf{n}}\cdot\boldsymbol{\beta})(\alpha\hat{\mathbf{n}} - 4\gamma\boldsymbol{\beta}) + \kappa(\dot{\alpha}\hat{\mathbf{n}} - 4\dot{\gamma}\boldsymbol{\beta} - 4\gamma\dot{\boldsymbol{\beta}})}{c\kappa^3 R} \right\}_{\text{ret}} . \qquad (11.2)$$

Appendix D presents the derivation of Eq. (11.2) from [4]. This retarded gravitational field is relativistically exact for all velocities and accelerations of the source mass, but is only valid in the weak-field approximation, in which $Gm / Rc^2 \ll \beta^2$. That is, Eq. (11.2) neglects terms of order $(Gm / Rc^2)^2$ and is only valid for gravitationally unbound quadrupoles.

With respect to Fig. 11.3 and Eq. (11.2), $\mathbf{u}$ is the 3- velocity of the source mass m; $\boldsymbol{\beta} = \mathbf{u} / c$ is the normalized 3-velocity; $\gamma = (1 - \beta^2)^{-1/2}$ is the Lorentz

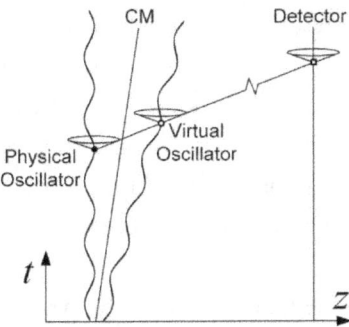

Fig. 11.2. World lines of physical and virtual oscillators, their center of mass 'CM', and detector from the model configuration of Fig. 11.1.

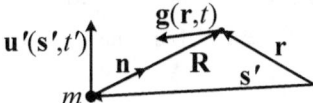

Fig. 11.3. Configuration of source mass m moving with retarded velocity $\mathbf{u}'(\mathbf{s}',t')$ and producing a field $\mathbf{g}(\mathbf{r},t)$ at a test particle at rest at $(\mathbf{r},t)$.

(relativistic) factor, and $\alpha \equiv 2\gamma - 1/\gamma$; $\mathbf{R} = \mathbf{r} - \mathbf{s}'$ is the displacement vector from the source position $\mathbf{s}'(t')$ to the test particle at $\mathbf{r}(t)$; $\mathbf{n} = \mathbf{R}/R$ is a unit vector; the factor $\kappa \equiv 1 - \mathbf{n} \cdot \boldsymbol{\beta}$ is the derivative with respect to t' of $t' + [R(t')/c] - t$; an overdot denotes differentiation with respect to t'; and the quantity in brackets $\{\ \}_{\text{ret}}$ and all primed quantities are to be evaluated at the retarded time $t' = t - R'/c$.

Just as does the retarded electric field of a point charge, the retarded gravitational (or inertial) field in Eq. (11.2), $\mathbf{g}(\mathbf{r},t)$, divides itself naturally into a 'velocity field', $\mathbf{g}_{\text{v}}(\mathbf{r},t)$, which is independent of acceleration and which varies as R^{-2}, and an 'acceleration field', $\mathbf{g}_{\text{a}}(\mathbf{r},t)$, which depends linearly on $\dot{\boldsymbol{\beta}}$ and which varies as R^{-1}. And just as for electromagnetic radiation, because of this range dependence, only the 'acceleration field' contributes to the dipole perturbation in the radiation zone, where $R/c \gg \beta/\dot{\beta}$.

Equation (11.2) is the retarded gravitational field of the mass m that would be observed by a distant unaccelerated observer to act on a test particle or a detector at rest at $(\mathbf{r},t)$. If the test particle moves, then the gravitational field measured by the moving test particle has additional 'gravimagnetic' terms. The following calculation applies only in the rest frame of the test particle and uses the impulse approximation, which has the test particle remaining at rest during the time the gravitational field of the moving particle acts upon it. (Any difference in fields caused by motion induced in the test particle by the source is of the same order as terms that have already been neglected in the weak-field approximation.)

In the source region and the near zone, the 'velocity field', which varies as R^{-2}, is generally more significant than the 'acceleration field'. For the special case of a source moving with constant velocity, the expression for the 'velocity field' in Eq. (11.2) was confirmed by an exact solution of Einstein's equation, valid even for strong fields, in [52, 53].

From Eq. (11.2), in the slow-velocity approximation for the source, to first order in β, the 'velocity field' of a mass m, measured at the stationary spacetime point $(\mathbf{r_0}, t)$ is

$$\mathbf{g}_v(\mathbf{r_0}, t) \approx -(Gm/R'^2)\{[1+2(\hat{\mathbf{n}} \cdot \boldsymbol{\beta})]\hat{\mathbf{n}} + \hat{\mathbf{n}} \times (\hat{\mathbf{n}} \times \boldsymbol{\beta})\}_{\text{ret}}, \qquad (11.3)$$

and the 'acceleration field' in the same slow-velocity approximation is

$$\mathbf{g}_a(\mathbf{r_0}, t) \approx -(Gm/cR')\{(\hat{\mathbf{n}} \cdot \dot{\boldsymbol{\beta}})[(1+3\hat{\mathbf{n}} \cdot \boldsymbol{\beta})\hat{\mathbf{n}} - 4\boldsymbol{\beta}] \\ +3\beta\dot{\beta}\hat{\mathbf{n}} - 4(1+2\hat{\mathbf{n}} \cdot \boldsymbol{\beta})\dot{\boldsymbol{\beta}}\}_{\text{ret}}. \qquad (11.4)$$

This slow-speed approximation of the weak field is valid only for $Gm/Rc^2 \ll \beta^2 \ll 1$. Equation (11.3) is written here as the sum of a parallel-polarized field in the (retarded) radial direction $\hat{\mathbf{n}}'$ and a transverse-polarized field perpendicular to $\hat{\mathbf{n}}'$. Equation (11.4) will be used next to calculate the dipole inertial radiation from the nonrelativistic unbound quadrupole in Fig. 11.1.

The model of an unbound quadrupole in Figs. 11.1 and 11.2 is equivalent to a physical oscillator at the origin and a virtual oscillator with identical mass, amplitude, and frequency, but π radians out of phase and moving at constant velocity ω/k in the z direction. That is, the position of the physical oscillator is $z = -a\cos\omega t$, and the position of the virtual oscillator is $z = b + (\omega t/k) + a\cos\omega t$, where $b \gg a$ is an initial displacement.

Although the momentum density of this system is time-dependent and non-periodic, the total momentum $m\omega / k$ is constant. The acceleration of the virtual oscillator, $c\dot{\boldsymbol{\beta}} = -a\omega^2(\cos\omega t)\hat{\mathbf{z}}$, is opposite that of the physical oscillator. But the velocity of the virtual oscillator, $c\boldsymbol{\beta} = [(\omega / k) - a\omega(\sin\omega t)]\hat{\mathbf{z}}$, contains a non-periodic (constant) term not shared by the physical oscillator.

Of course, the virtual oscillator representing the propagating stress wave in the rod can just as well be a physical oscillator of opposite phase and separating at constant speed ω / k from the oscillator at the origin. In either case the momentum of the system is constant. But the physical grounding of the model as shown in Fig. 11.1 is more readily apparent.

Averaged over oscillations, the retarded separation of the two oscillators is taken to be $s' = b + \omega t' / k$, where the initial separation is $b \gg a$, and the speed of the virtual oscillator is $\omega / k \ll c$. Neglecting terms of order s' / R', except in the argument of $\cos\omega t'$, the retarded range from both oscillators to the observation point, $R' \approx r_0$, is nearly constant, as is the direction of both, $\hat{\mathbf{n}}' \approx \hat{\mathbf{n}}$. The retarded time of the physical oscillator is $t'_0 \approx t - r_0 / c$, and the retarded time of the virtual oscillator is $t'_s \approx [1 + (\omega / kc)\cos\theta][t - (r_0 - b\cos\theta) / c]$, where $\cos\theta \equiv \hat{\mathbf{n}} \cdot \hat{\mathbf{z}}$. The factor $1 + (\omega / kc)\cos\theta$ represents a Doppler frequency shift, and the factor $t - (r_0 - b\cos\theta_0) / c$ represents a phase difference of the retarded time of the virtual oscillator with respect to the retarded time of the physical oscillator.

From Eq. (11.4) to zeroth order in β, the combined field perturbation of the two oscillators in the radiation zone is

$$\mathbf{g_a}(\mathbf{r}_0, t) \approx (3Gma\omega^3 / kr_0 c^3)$$
$$\times \left\{ [(1 + 3\cos^2\theta)\hat{\mathbf{n}}_\| + 4\sin\theta\cos\theta\hat{\mathbf{n}}_\perp]\cos\omega(t - r_0 / c) \right. \qquad (11.5)$$
$$\left. + ks'[\cos^2\theta\hat{\mathbf{n}}_\| - (4/3)\sin\theta\cos\theta\hat{\mathbf{n}}_\perp]\sin\omega(t - r_0 / c) \right\}$$

Here, $\hat{\mathbf{n}}_\| \equiv \hat{\mathbf{n}}$ is the unit vector in the direction parallel to propagation of the wave, and $\hat{\mathbf{n}}_\perp \equiv \hat{\mathbf{n}} \times (\hat{\mathbf{n}} \times \hat{\mathbf{z}}) / \sin\theta$ is a unit vector in the transverse direction, as shown in Fig. 11.1.

From the weak-field Lagrangian density, the energy density of a gravity wave in the radiation zone is the t^{00} component of the canonical energy-momentum tensor $t^{\mu\nu}$, which is $t^{00} \approx g^2 / 8\pi G$ [54]. The energy flux is the energy density times c, and the power radiated per unit solid angle is

$dP / d\Omega \approx cg^2 r^2 / 8\pi G$. At zero oscillator separation speed, that is, for $\omega / k = 0$ and $s' = b$, corresponding to a matched pair of harmonic oscillators with opposite phase, the angular distribution of root-mean-square (rms) dipole inertial power is

$$\overline{dP} / d\Omega \approx P_0 (9\cos^4 \theta + 16\sin^2 \theta \cos^2 \theta) / 16\pi , \tag{11.6}$$

where $P_0 \equiv Gm^2 a^2 b^2 \omega^6 / c^5$. In the other limit, $\omega / k \gg \omega s'$, in which the Doppler shift dominates the phase difference, the angular distribution of rms power is

$$\overline{dP} / d\Omega \approx (P_0 / 16\pi k^2 b^2)[9(1 - 3\cos^2 \theta)^2 + (12\sin \theta \cos \theta)^2]. \tag{11.7}$$

The amplitude of the dipole inertial wave in Eq. (11.5) is proportional to the acceleration $a\omega^2$, and has terms proportional to the separation s' and to the separation speed ω / k of the oscillators. Unlike classical quadrupole radiation, or electromagnetic radiation for that matter, the polarization of dipole inertial waves is not purely transverse in the radiation zone. Figure 11.4 shows the angular distribution of dipole power for the parallel-polarized and transverse-polarized components of the inertial wave in Eq. (11.5) in the two limits, $ks' \gg 1$ from Eq. (11.6) and $ks' \ll 1$ from Eq. (11.7).

By integrating Eq. (11.6) over all solid angles, the dipole inertial-wave rms power is found to be $\overline{P} \approx 59 P_0 / 60$ for $ks' \gg 1$, of which $27 P_0 / 60$ is parallel polarized. From Eq. (11.7), it is found to be $\overline{P} \approx 33 P_0 / 5k^2 b^2 \gg P_0$ for $ks' \ll 1$,

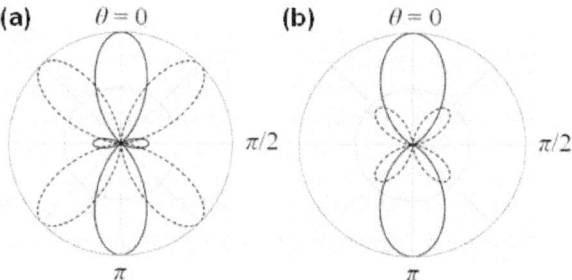

Fig. 11.4. Polar plots of angular distributions of parallel-polarized (solid curves) and transverse-polarized (dashed curves) dipole power, $dP / d\Omega$, from configuration shown in Fig. 11.1, for: (a) constant oscillator separation; and (b) $\omega / k \gg \omega s'$.

of which $9P_0 / 5k^2b^2$ is parallel polarized.

If the separation speed ω / k of the two oscillators is slow, the dipole power in the radiation zone of the system shown in Fig. 11.1 is of the same quadrupolar order, $\sim P_0$, as the classical quadrupole power of a gravitationally bound quadrupole. Although the gravity-wave power of two unbound, slowly-separating dipoles is of quadrupolar order, it is of dipolar character. Classical (transverse traceless) gravity waves from a gravitationally bound quadrupole have polarizations that are tensors of rank 2. Since transverse traceless gravity waves do not accelerate a lone particle, they can be detected only by the *tidal force acting on a quadrupole detector*. Their signal strengths are many orders of magnitude lower than those of dipole inertial waves, which have vector polarizations and can be detected by the *force acting directly on a dipole detector*.

Chapter 12
Dipole Inertial Radiation, Inertial Forces, and Newton's Third Law

Einstein observed in his 1913 letter to Mach [1, 2], "If one accelerates a heavy shell of matter S, then a mass enclosed by that shell experiences an accelerative force." This section calculates, within the weak-field, slow-velocity approximation of general relativity, the inductive inertial effect to which Einstein was referring. The linearized field equations of Ch. 11 give the same result for the inverse effect: If one accelerates a mass, then a shell of matter enclosing that mass experiences an accelerative force. The inductive inertial forces are the same whether an accelerating shell is acting on a mass at its center or an accelerating mass at the center is acting on a shell surrounding it.

The calculations for the latter case, an accelerating mass at the center, are shown here because these calculations are more suggestive of an origin of Newton's third law. As in the example considered in Einstein's letter and in Ch. 11 of this book, the accelerating mass in this example is a gravitationally unbound quadrupole. And as in Einstein's letter, neither dipole moment nor momentum is explicitly conserved in this example. The calculation is instructive nonetheless.

As described in Ch. 11, the dipole inertial field of an accelerating mass interferes with the corresponding field of the mass that is producing the acceleration to determine the angular distribution of power that is transported to the radiation zone. But if the mass causing the acceleration is of sufficiently different character than the mass that is being accelerated, then there could be virtually no interference of their corresponding inertial fields in the radiation zone. For example, suppose a passing electromagnetic wave accelerates a solitary charge. The net momentum of the accelerating charge and the electromagnetic field disturbance produced by the charge is zero, but the dipole inertial fields of each will have virtually no interference in the radiation zone.

The calculation of dipole inertial radiation produced by a solitary mass in accelerated motion is instructive because of the insight it gives to Newton's third law, "To every action there is always opposed an equal reaction; or, the mutual actions of two bodies upon each other are always equal, and directed to contrary parts." This section shows that the dipole inertial radiation produced by an accelerating mass carries momentum and power to the radiation zone that suggests a relationship with an inertial force acting instantaneously to oppose the acceleration. That is, dipole inertial radiation may be the mechanism that ensures that masses do not spontaneously accelerate with no applied force. The conjecture is that the inertial force produced by inertial radiation would instantaneously counter, in magnitude and direction, any such spontaneous acceleration.

When a mass is accelerated by a force, the mass does work against the local inertial field of the universe and emits dipole fields into the source region that depend only on the velocity and acceleration of the mass, and not on what caused the acceleration. Just as some or all of the dipole electromagnetic fields emitted by an accelerated charge might be reabsorbed by other charges in the source region [50], some or all of the dipole fields of an accelerated mass might be reabsorbed by other masses in the source region. If the dipole moment of a system of nonrelativistic masses in the source region is constant, then the inertial radiation will generally be only of quadrupole order. But if the sources of the dipole fields are relativistic, or of very different character from each other, or have large differences in velocity or in range to a detector, then the inertial radiation at a detector in the radiation zone will generally be of dipole order. The following calculation of the dipole inertial radiation produced by a solitary mass in accelerated motion assumes no interference by fields of other masses. The calculation is intended to reproduce quantitatively the qualitative result reported by [1, 2] in his letter to Mach.

As shown in Fig. 12.1, consider a thin spherical shell of mass M_0, radius R_0, and uniform mass per unit area. With reference to the coordinates and vectors in Fig. 12.1, let the instantaneous location of the particle of mass m at the retarded time t' be the origin, and the instantaneous acceleration and normalized velocity of the mass be $\mathbf{a}(t') = a(t')\hat{\mathbf{z}}$ and $\boldsymbol{\beta}(t') = \beta(t')\hat{\mathbf{z}}$, respectively.

From Eq. (11.3), to first order in β, the 'velocity field' of the mass m at the stationary spacetime point $(\mathbf{r}_0, t)$ on the surface of the spherical shell, where $r_0 = R_0$, is

$$\mathbf{g}_v(\mathbf{r}_0, t) \approx -(Gm / R_0^2)\left[(1 + 2\beta\cos\theta)\hat{\mathbf{n}}_\parallel + (\beta\sin\theta)\hat{\mathbf{n}}_\perp\right]_{\text{ret}}, \tag{12.1}$$

and from Eq. (11.4), the 'acceleration field' at the spherical shell in the same

96

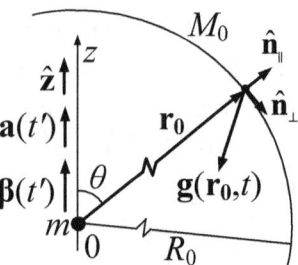

Fig. 12.1. Model configuration for calculating inertial radiation of an accelerating mass at a spherical shell in the radiation zone at $(\mathbf{r}_0, t)$.

slow-velocity approximation, to first order in β, is

$$\mathbf{g_a}(\mathbf{r}_0, t) \approx (Gma/c^2 R_0)\{3(\cos\theta - \beta + 3\beta\cos^2\theta)\hat{\mathbf{n}}_{\parallel} \\ +4\sin\theta(1+3\beta\cos\theta)\hat{\mathbf{n}}_{\perp}\}_{\text{ret}} \qquad (12.2)$$

When averaged over the spherical shell, the only nonzero components of the 'velocity field' and 'acceleration field' by symmetry can be the z components,

$$\mathbf{g_v} \cdot \hat{\mathbf{z}} \approx -(Gm/R_0^2)\{\cos\theta - \beta + 3\beta\cos^2\theta\}_{\text{ret}}, \qquad (12.3)$$

$$\mathbf{g_a} \cdot \hat{\mathbf{z}} \approx +(Gma/c^2 R_0)\{3 + \sin^2\theta + 9\beta\cos\theta(1+\cos^2\theta)\}_{\text{ret}}. \qquad (12.4)$$

To first order in β, the average of the 'velocity field' over the spherical shell is zero. To first order in β, the average of the 'acceleration field' over the spherical shell is

$$\oiint (d\Omega/4\pi)(\mathbf{g_a} \cdot \hat{\mathbf{z}})\hat{\mathbf{z}} \approx +(11G/3c^2 R_0)m\mathbf{a}, \qquad (12.5)$$

where $d\Omega = d\phi\sin\theta d\theta$ is an element of solid angle on the spherical shell. Thus, the averages over a shell of both the 'velocity field' and 'acceleration field' of a mass accelerating at the center are independent of β to first order in β, and the average force exerted on the spherical shell of mass M_0 by the mass m is

$$M_0 \mathbf{g_a} \approx \{11GM_0 / 3c^2 R_0\}m\mathbf{a}.$$ (12.6)

In the inverse effect, the average force exerted on the mass m by a spherical shell of mass M_0 with an acceleration $\mathbf{a}$ is identical to the force in Eq. (12.6).

Suppose the particle of mass m at rest at the center of a surrounding shell of mass M_0 undergoes a constant acceleration $\mathbf{a}$ for a brief time Δt, so that it acquires a momentum $\Delta\mathbf{p} = m\mathbf{a}\Delta t$. From Eq. (12.2), the specific momentum imparted to the spherical shell to zeroth order in β is

$$\Delta\mathbf{V_0}(\theta) \approx +\left(G\Delta p / c^2 R_0\right)\left(3\cos\theta\hat{\mathbf{n}}_{\parallel} - 4\sin\theta\hat{\mathbf{n}}_{\perp}\right).$$ (12.7)

The average specific impulse imparted to the spherical shell is

$$\Delta\bar{\mathbf{V}}_0 \equiv \oiint(d\Omega / 4\pi)\Delta\mathbf{V_0}(\theta) \approx +11G\Delta\mathbf{p} / 3c^2 R_0.$$ (12.8)

From Eq. (12.2), the radiated dipole power per unit solid angle at the spherical shell is

$$dP / d\Omega \approx (cr^2 / 8\pi G)\left|\mathbf{g_a}(\mathbf{r_0},t)\right|^2$$
$$\approx (2Gm^2 a^2 / \pi c^3)\{1-(7/16)\cos^2\theta\}_{\text{ret}}$$ (12.9)

From integrating Eq. (12.9) over solid angle, the total dipole power radiated to the spherical shell in the radiation zone by the accelerating mass is

$$P \approx (25\pi / 8)Gm^2 a^2 / c^3.$$(12.10)

The energy in the pulse radiated to the spherical shell is

$$P\Delta t \approx (25\pi / 8)Gma\Delta p / c^3.$$ (12.11)

The total kinetic energy imparted to the spherical shell is

$$\oiint(d\Omega / 4\pi)M_0[\Delta\mathbf{V_0}(\theta)]^2 / 2 \approx (41M_0 / 6)(G\Delta p / c^2 R_0)^2.$$ (12.12)

If M_0 and R_0 are of the order of the mass and radius of the universe, then the quantity in the brackets in Eq. (12.6) is of the order of the 'sum for inertia', that is, of the order of 1. If this calculation were done exactly, without the linearization and other approximations of this simple spherical-shell model, one might reasonably expect that the force exerted on the rest of the mass of the universe by the mass m undergoing an acceleration $\mathbf{a}$ would be exactly $m\mathbf{a}$, and that this force is conveyed to the rest of the mass of the universe, in the future light cone of the mass m, by the inertial radiation emitted during acceleration. If so, then from Eqs. (12.11) and (12.12), the ratio of kinetic energy imparted to the mass of the universe to the energy carried in the radiation pulse is of order $0.2c\Delta t / R_0 \ll 1$.

To summarize how this simple model may relate to Newton's third law: An accelerating particle does work on the local inertial field of the universe. The work on the local field results in the particle emitting dipole fields that transmit momentum and energy to the radiation zone, subject to interference by fields emitted by other masses in the source region. In reaction to the emission of field momentum by the particle, and for conservation of momentum, an impulse is delivered to the particle in a direction opposing the acceleration. The impulse that will ultimately be delivered by the inertial radiation of the accelerating particle to the mass in the future light cone of the particle may be exactly equal and opposite to the impulse delivered by the inertial field to the accelerating particle, which could be the equal and opposite reaction of Newton's third law.

Of course, to be applicable over cosmological ranges, any model must employ the full nonlinearity of the field equations of general relativity. The calculation here of dipole inertial radiation of an isolated unbound quadrupole is based on the linearized equations and is not applicable over cosmological ranges. The calculation is suggestive nonetheless.

For accelerations less than or about the inertial drag constant D, the 'velocity field' can be greater than the 'acceleration field' even in the radiation zone. In the model representing the universe as a thin spherical shell, the power radiated per unit solid angle at the shell by an isolated accelerating particle, from Ch. 11, is

$$dP / d\Omega \approx \left(cR_0^2 / 8\pi G \right) \left[\mathbf{g}_v(\mathbf{r}_0, t) + \mathbf{g}_a(\mathbf{r}_0, t) \right]^2. \tag{12.13}$$

Using the fields from Eqs. (12.1) and (12.2) of a particle instantaneously at rest, but with acceleration $\mathbf{a}(t') = a(t')\hat{\mathbf{z}}$, the angular distribution of parallel-polarized power radiated to the shell is

$$\frac{dP_{\parallel}}{d\Omega} \approx \frac{Gm^2}{8\pi c^3}\left(3a\cos\theta - c^2 / R_0\right)^2 , \tag{12.14}$$

and the angular distribution of transverse-polarized power radiated to the shell is

$$\frac{dP_{\perp}}{d\Omega} \approx \frac{Gm^2}{8\pi c^3}\left(4a\sin\theta\right)^2 . \tag{12.15}$$

The angular distribution of total power, parallel and transverse-polarized, radiated to the shell is

$$\frac{dP}{d\Omega} \approx \frac{Gm^2}{8\pi c^3}\left[16a^2 - 7a^2\cos^2\theta - \left(6ac^2 / R_0\right)\cos\theta\right]. \tag{12.16}$$

Because the 'velocity field' varies as R^{-2}, and the 'acceleration field' varies as R^{-1}, generally only the 'acceleration field' contributes to the dipole perturbation in the radiation zone at large R. Equations (12.1) and (12.2), however, show that the ratio of 'acceleration field' to 'velocity field' in the simple model calculated here scales as $g_a / g_v \sim aR/c^2$. If the range R is taken to be of the order of the radius of the universe, then the ratio of 'acceleration field' to 'velocity field' scales as $g_a / g_v \sim a/D$, where $D \sim c^2 / R$ is the inertial drag constant. That is, the 'acceleration field' of a particle dominates at cosmological ranges unless the particle acceleration is less than or the order of D. In that event, the 'velocity field' may exceed the 'acceleration field' at cosmological ranges. But in that event, Eq. (12.16) shows that the angular distribution of power, $dP / d\Omega$, is less than or the order of $Gm^2 D^2 / c^3$ for a nonrelativistic particle decelerated only by inertial drag, and is therefore negligibly small, given the approximations of this analysis.

Chapter 13
Dipole Inertial Noise and the Schrödinger Wave Equation

Dipole inertial noise can account classically for all the quantum mechanical effects described by the time-dependent Schrödinger wave equation, if the spectral density of the inertial field is about $l_p^2 \omega^3$, where $l_p \equiv (\hbar G / c^3)^{1/2}$ is the Planck length and ω is the angular frequency of the noise, and if the noise spectrum is cut off at about the Planck frequency, $\omega_p \equiv (c^5 / \hbar G)^{1/2}$.

Nelson [55] has shown that "any particle [of mass m] ... that is constantly undergoing a Brownian motion with diffusion coefficient $\hbar / 2m$ obeys the Schrödinger equation," where $\hbar = h / 2\pi$ is the reduced Planck constant. Therefore, any field that causes all particles constantly to undergo such a Brownian motion may be taken to be the cause of all the quantum mechanical effects arising from the Schrödinger equation.

One such field that could cause *all* particles, charged and uncharged, bound and unbound, constantly to undergo such a Brownian motion is a uniform background of *dipole* inertial noise. Classical (transverse traceless) gravity waves are quadrupolar and do not cause isolated particles to move. Dipole inertial waves, on the other hand, can cause isolated particles to move by pushing and pulling them in the direction of polarization.

Chapter 11 showed that dipole inertial disturbances from gravitationally unbound mass quadrupoles propagate to the radiation zone without complete destructive interference, as long as the accelerating dipole constituents of the mass quadrupoles are either unbound or bound by some force other than gravity [9]. Examples treated in [9] of such gravitationally unbound quadrupoles that produce dipole inertial waves in the radiation zone include mechanical oscillators and rotors, particle storage rings, and powerful astrophysical events, like asymmetric supernovas. The power spectral density, angular distribution, and polarization of dipole inertial radiation are calculated in [9], as are absorption and scattering cross sections for plane dipole inertial waves incident on simple mass oscillators.

Since [9] calculates inertial fields of particles with arbitrary relativistic velocity, one can confirm the existence of dipole inertial radiation even in the slow-velocity and weak-field approximations of general relativity, as was demonstrated in Ch. 11. Since sources of dipole inertial waves are ubiquitous in the universe, and since any dipole inertial wave incident on an otherwise isolated system of masses should be expected to change the dipole moment of the system and to be scattered by the system, a stochastic background of dipole inertial noise is expected to exist. In this section, the Wiener-Khinchine relations and Nyquist's theorem are used to estimate the spectral density of dipole inertial noise needed to produce Brownian motion of all particles with diffusion coefficient $\hbar / 2m$. This spectral density is found to be about $l_p^2 \omega^3$.

Einstein famously wrote [56], "I, in any case, am convinced that He does not play dice," often popularized as, 'God does not play dice with the universe.' If Einstein's intuition on causality versus randomness was correct, and if the quantum mechanical effects described by the Schrödinger equation do arise from a physical noise background, then it might even be said that the Schrödinger equation offers a precise diagnostic of the spectral density of this background.

Inertial noise exists everywhere and is screened nowhere. Inertial noise is the seemingly random fluctuation about a mean local gravitational field produced by the motion of mass on the past light cone. At every spacetime point $(\mathbf{x},t)$, the inertial field $\mathbf{g}(\mathbf{x},t)$ of the mass on the past light cone of that point is a highly dynamical function of time. A relativistically exact equation for the weak retarded gravitational field of a particle in arbitrary relativistic motion is given in Eq. (11.2). (And an exact equation for the strong retarded gravitational field of a spherical mass in uniform motion was derived in [52].) If the expression in Eq. (11.2), corrected for cosmological and inertial and strong-field effects, could be integrated over the mass on the past light cone of $(\mathbf{x},t)$, then to the extent that the fields add linearly, the inertial noise at $(\mathbf{x},t)$ could in principle, though not in practice, be calculated deterministically.

Of course, a fluctuating inertial field produces fluctuations of particle motion, much like Brownian motion of a particle in a fluid. However weak it may be, inertial noise must induce Brownian motion with some diffusion coefficient in all particles. These inertial fluctuations must act on a particle oscillator just as thermal fluctuations of voltage act on an oscillating electrical circuit.

Since inertial fluctuations act on particle oscillators everywhere, a corresponding mechanism must also act everywhere to dissipate the noise energy from the oscillators. Otherwise, the resonant Fourier frequency component of the fluctuations would drive an *undamped* oscillator to failure. (Over any given time period, each Fourier component of a fluctuating force is a sinusoidal force of constant amplitude. The resonant Fourier components would make the amplitude of an *undamped* harmonic oscillator grow exponentially without bound.) An equilibrium of a fluctuating driving force with energy dissipation in any oscillating system will result in a ground-state or 'zero-point' energy of the oscillating system.

Consistent with the finding by [55] that any particle that is constantly undergoing a Brownian motion with diffusion coefficient $\hbar/2m$ obeys the Schrödinger equation is that the Heisenberg uncertainty principle for position and momentum, which is one result of the Schrödinger equation, may be written as $\Delta x (\Delta x / \Delta t) \geq \hbar / 2m$. It follows that if some physical mechanism accounts for a diffusion coefficient of $\hbar/2m$ for all particles, then the Schrödinger equation may be considered a consequence of that physical mechanism. Similarly, if some physical mechanism accounts for a ground-state energy of $\hbar \omega_0 / 2$ for all (one-dimensional) harmonic oscillators with ground-state angular frequency ω_0, then the Schrödinger equation may be considered a consequence of that physical mechanism.

Dipole inertial noise may be the physical mechanism that accounts for a ground-state energy of one-dimensional harmonic oscillators of $\hbar \omega_0 / 2$ and a diffusion coefficient of $\hbar/2m$ for all particles. Just as the Schrödinger wave equation does, inertial forces apply equally to charged and uncharged particles. No particles are shielded from inertia. The apparent constancy of the ground-state mode energy (for hydrogen atoms, say) throughout the observable universe, moreover, suggests that the fluctuating driving force could arise from the homogeneous distribution of mass and the uniformity of the inertial field throughout the universe.

As explained in Ch. 11, dipole inertial perturbations can be produced and can propagate into the far (radiation) zone even from 'closed systems' in which the mass dipole moment is zero. A simple example of a gravitationally unbound quadrupole with zero dipole moment that produces dipole inertial waves in the radiation zone is the oscillator system shown in Fig. 11.1. In the limit in which the velocity of the stress wave goes to zero, the system in Fig. 11.1 comprises two identical oscillators with constant separation d and opposite phase. Each oscillator has mass m, amplitude a_0, and angular frequency ω_0. In the radiation zone, in which $a_0 \ll d \ll c/\omega_0 \ll r$, and in the slow-velocity, weak-field approximation of general relativity, in which $Gm/dc^2 \ll \beta^2 \ll 1$, the dipole inertial field radiated by just the one oscillator at the origin, from Eq. (11.2) is

$$\mathbf{g_0}(\mathbf{r},t) \approx \left(Gma_0\omega_0^2/rc^2\right)\left(\cos\theta\,\hat{\mathbf{n}}_{\parallel} - 4\hat{\mathbf{z}}\right)\cos\omega_0 t_0', \tag{13.1}$$

where $t_0' = t - r/c$ is the retarded time of the mass oscillating about the origin. Equation (13.1) represents the inertial field in the radiation zone of an isolated particle undergoing harmonic oscillations at the origin, driven for example by an electromagnetic or an inertial wave.

The angular distribution of rms dipole power from the inertial field in Eq. (13.1) of the isolated dipole oscillator is

$$\frac{d\bar{P}}{d\Omega} \approx \frac{cg^2 r^2}{16\pi G} \approx \frac{1}{\pi}\left[1 - (7/16)\cos^2\theta\right]P_d, \tag{13.2}$$

where $P_d \equiv G(ma_0\omega_0^2)^2/c^3$. Integrating Eq. (13.2) over solid angle gives the total dipole rms power radiated by a solitary oscillating mass of amplitude a_0 and angular frequency ω_0 as

$$\bar{P} \approx (41/12)P_d. \tag{13.3}$$

This expression for radiated power is the inertial radiation equivalent of the electromagnetic Larmor power radiated by an isolated electric dipole oscillator.

To calculate the diffusion coefficient of the inertial noise field, consider an isolated harmonic oscillator in equilibrium with the inertial noise field. That is, the oscillator is at an equilibrium ground-state energy driven by the fluctuations of the inertial noise field and damped by the dipole inertial power radiated by the oscillator, with magnitude given by Eq. (13.3).

A dipole inertial wave at an oscillator is a linear superposition of plane-wave Fourier components having constant field amplitude. Consider the oscillator to be a simple mass quadrupole, comprising an oscillating mass of natural angular velocity ω_0 bound with an isotropic restoring force and damping constant Γ to a much heavier stationary structure.

The response $x(t)$ of a one-dimensional harmonic oscillator of natural frequency ω_0, driven by a random forcing function $g(t)$, and damped with decay constant Γ, obeys

$$d^2x / dt^2 + \Gamma \, dx / dt + \omega_0^2 x = g(t), \tag{13.4}$$

where $x(t)$ represents the displacement of a particle oscillator of mass m, and $g(t)$ represents the inertial noise field.

Appendix E discusses the general statistical properties of a noise field such as the inertial noise field $g(t)$. The calculation considers a large number N of field modes within a specified frequency band, the modes having random phase and some distribution of amplitudes, which could be random. The appendix considers a plane-wave field,

$$g(z,t) = \sum_{i=1}^{N} g_i \cos\left[\omega_i (t - z / c) + \phi_i\right], \tag{13.5}$$

propagating with velocity c in the z direction and comprising N field modes, where g_i is the amplitude, ω_i is the angular frequency, and ϕ_i is the phase of the ith field mode. The appendix shows that, as long as N is very large and the frequency band is narrow, then the field modes within the frequency band combine to produce effectively a single mode with a frequency $\bar{\omega}$ about equal to the mean frequency or central frequency of the band and with an amplitude $g_0 \approx N^{1/2} \bar{g}_i$, where $\bar{g}_i$ is the mean amplitude of each of the modes in the band. That is, within the frequency band, the driving field at a particular location, such as an oscillator, is

$$g(t) = g_0 \cos(\bar{\omega}t + \phi_0). \tag{13.6}$$

When this expression for driving field amplitude is used in the damped, driven harmonic oscillator equation, Eq. (13.4), the amplitude of the oscillations as a function of the driving frequency, $\omega = \bar{\omega}$, found from the solution of Eq. (13.4), is

$$a(\omega) = (g_0 / \Gamma\omega_0)\left[1 + (\omega_0^2 - \omega^2)^2 / (\Gamma\omega_0)^2\right]^{-1/2}. \tag{13.7}$$

Figure 13.1 shows (for $\omega_0 = 20\Gamma$) the amplitude of the oscillator as a function of driving frequency ω. The resonant amplitude at ω_0 is $a_0 = g_0 / \Gamma\omega_0$. The full width of the resonance in $a(\omega)$ at half-maximum-amplitude is $3^{1/2}\Gamma$.

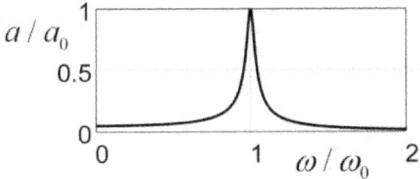

Fig. 13.1. Amplitude normalized to resonant amplitude $a_0 = g_0 / \Gamma \omega_0$ vs. driving frequency normalized to resonant frequency ω_0.

The solution in Eq. (13.7) applies to each of the field modes individually as well as to the combined driving field amplitude. Appendix E shows that, as long as N is very large and the frequency band is narrow, then the oscillation amplitudes a_i of each mode within the frequency band combine to produce effectively a single amplitude $\bar{a} \approx N^{1/2} \bar{a}_i$, where $\bar{a}_i$ is the mean amplitude of each of the modes in the band. That is, within the frequency band, the displacement of an oscillator is

$$x(t) = \bar{a} \cos(\bar{\omega} t + \theta_0).$$ \hfill (13.8)

Only the noise field modes within about a line width Γ of the resonant frequency ω_0 contribute significantly to the displacement of the oscillator. The driving field within a line width of ω_0 is of order g_0. The amplitude of the oscillator within a line width of ω_0 is of order $\bar{a} \sim g_0 / \Gamma \omega_0$. The velocity of the oscillator within a line width of ω_0 is of order $\bar{a} \omega_0 \sim g_0 / \Gamma$. The diffusion coefficient $\Delta_d = \bar{a}^2 \omega_0$ within a line width of ω_0 is of order $\Delta_d \sim g_0^2 / \Gamma^2 \omega_0$. The in-band driving field amplitude, g_0, therefore scales as

$$g_0^2 \sim \Delta_d \Gamma^2 \omega_0.$$ \hfill (13.9)

The dipole power radiated by the oscillator, $\bar{P}$, from Eq. (13.3) scales as the energy of the oscillator times the decay constant, that is, $G(m\bar{a}\omega_0^2)^2 / c^3 \sim m\bar{a}^2 \omega_0^2 \Gamma$. The decay constant for dipole radiation from an oscillator, therefore, scales as

$$\Gamma \sim Gm\omega^2 / c^3.$$ \hfill (13.10)

A simple mode density calculation shows that the spectral density $\rho(\omega)$ of inertial noise must be proportional to ω^3. In a cubic volume of length L, with periodic boundary conditions on the inertial field, the Cartesian components of the wavenumber 3-vector $\mathbf{k}$ are

$$k_x = 2\pi l / L, \quad k_y = 2\pi m / L, \quad k_z = 2\pi n / L, \tag{13.11}$$

where l, m, and n are integers. The wavenumber is related to angular frequency by

$$\omega^2 = (k_x^2 + k_y^2 + k_z^2)c^2. \tag{13.12}$$

From Eq. (13.11), the volume of each mode in $\mathbf{k}$ space is

$$dk_x dk_y dk_z = (2\pi / L)^3. \tag{13.13}$$

Since the total volume in $\mathbf{k}$ space of wavenumber vectors with magnitudes between 0 and k is $4\pi k^3 / 3$, from Eq. (13.13), the total number of modes with magnitudes between 0 and k is

$$N_k = n_p k^3 L^3 / 6\pi^2, \tag{13.14}$$

where n_p is the number of possible directions of polarization. The total number of modes per unit volume between $\omega = 0$ and ω is

$$n_\omega \equiv N_k / L^3 = n_p \omega^3 / 6\pi^2 c^3. \tag{13.15}$$

The mode density per unit angular frequency is

$$dn_\omega / d\omega = n_p \omega^2 / 2\pi^2 c^3, \tag{13.16}$$

the same as it is for an electromagnetic field.

As long as the frequency band is narrow, the energy density per unit angular frequency within a line width of the resonant frequency ω_0 scales as

$$\bar{\varepsilon}_0 dn_\omega / d\omega \sim \bar{\varepsilon}_0 \omega_0^2 / c^3, \tag{13.17}$$

where $\bar{\varepsilon}_0$ is the average energy per in-band inertial-radiation mode. The in-band energy density then scales as

$$\Gamma \overline{\varepsilon}_0 dn_\omega / d\omega \sim \overline{\varepsilon}_0 Gm\omega_0^4 / c^6 , \tag{13.18}$$

where the scaling of the decay constant from Eq. (13.10) was used. But the in-band energy density of radiation is also given by $g_0^2 / 8\pi G$ from Eq. (13.2) and Ch. 11. The in-band driving field amplitude, g_0, therefore scales as

$$g_0^2 \sim G^2 m\omega_0^4 \overline{\varepsilon}_0 / c^6 . \tag{13.19}$$

Comparing Eqs. (13.9) and (13.19) shows that the diffusion coefficient scales as

$$\Delta_d \sim \overline{\varepsilon}_0 / m\omega_0 . \tag{13.20}$$

In the reference frame in which the mean field vanishes at a point, that is, where $\langle g(t) \rangle = 0$, the mean square of the fluctuating noise field, $\langle g^2(t) \rangle$, is equal to its dispersion. From the Wiener-Khintchine relations, the spectral density $\rho(\omega)$ of the inertial noise field is defined in terms of its dispersion by

$$\langle g^2(t) \rangle = 2\int_0^\infty \rho(\omega)d\omega . \tag{13.21}$$

From Eq. (13.17) then, the energy density per unit angular frequency within a line width of the resonant frequency ω_0 scales as $\overline{\varepsilon}_0 dn_\omega / d\omega \sim \rho(\omega_0)/G$ $\sim \overline{\varepsilon}_0 \omega_0^2 / c^3$, and the spectral density scales as

$$\rho(\omega) \sim \overline{\varepsilon} G\omega^2 / c^3 . \tag{13.22}$$

Although the damping-force approximation in Eq. (13.10) applies only for $\Gamma \ll \omega_0$, the inertial dipole radiation spectrum of an oscillator nevertheless must be cut off at an angular frequency $\omega_c \sim \Gamma$, at which an energy of the order of the rest energy of the oscillator is radiated away in the order of one period. The dispersion of the inertial noise field, $\langle g^2(t) \rangle$, the spectral density of inertial noise, $\rho(\omega)$, and its cut-off frequency, ω_c, must all be independent of the characteristics m, ω_0, and $\overline{a}$ of any particular oscillator. The cut-off frequency can only depend on the natural constants $\hbar$, G, and c. The only combination of such constants that satisfies this condition is $\omega_c^2 \approx c^5 / \hbar G$. That is, the cut-off frequency is of the order of the Planck frequency. Sakharov [57] used the Planck frequency as the maximum cutoff frequency in a related calculation. By similar dimensional analysis, the dispersion of the inertial noise field scales as

$$\left\langle g^2(t) \right\rangle \sim \omega_p^{\,2} c^2 , \tag{13.23}$$

and the highest acceleration of a particle oscillator is of order $\omega_p c$, that is, acceleration to relativistic speed in a single wave cycle. Since the mode structure and mode density is the same for the inertial field as for the electromagnetic field, from Eqs. (13.22) and (13.23), the spectral density of the inertial noise field scales as

$$\rho(\omega) \sim \hbar G \omega^3 / c^3 \sim l_p^{\,2} \omega^3 . \tag{13.24}$$

If the spectral density of the inertial noise field scales as ω^3 with a cutoff roughly at ω_p, as shown in Fig. 13.2, then the inertial noise field could account for a diffusion coefficient for all particles of $\hbar / 2m$, and therefore, according to [55], may be taken to be cause of all the quantum mechanical effects arising from the Schrödinger wave equation.

Nyquist's theorem helps explain why the spectral density of the inertial noise field should have a cutoff frequency, and why the cutoff frequency should be of the order of ω_p. Nyquist's theorem is just a mathematical statement that only those Fourier frequency modes of a random force within a line width Γ about a resonant frequency ω_0 contribute significantly to the effective driving force on an oscillator. Appendix E shows that although these Fourier frequency modes have random phase, their linear superposition over any time that is short compared to their coherence time Γ^{-1} is a simple sinusoidal force of fixed phase at the resonant frequency. Nyquist's theorem relates this effective sinusoidal force to the decay constant in Eq. (13.10) through the line width Γ of the frequency resonance.

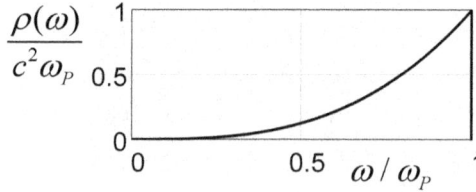

Fig. 13.2. Normalized spectral density of inertial noise field *vs.* normalized angular frequency with cutoff at Planck frequency that could account for effects of Schrödinger equation.

General statistical arguments applied to a damped harmonic oscillator in a random force field require that the mean kinetic energy of the oscillator be a constant $E_0/2$. Using Nyquist's theorem to evaluate the integral in the solution to Eq. (13.4), we find

$$E_0/2 = m\left\langle (dx/dt)^2 \right\rangle / 2 = m\int_0^\infty \frac{\rho(\omega)d\omega}{\left| \Gamma + i(\omega^2 - \omega_0^2)/\omega \right|^2} = \frac{\pi m \rho(\omega_0)}{2\Gamma(\omega_0)}. \qquad (13.25)$$

As expected, Eqs. (13.10), (13.24), and (13.25) show that the mean energy of the oscillator scales as

$$E_0 \sim \hbar\omega_0. \qquad (13.26)$$

If an elementary particle is driven at the Planck frequency for even a single wave cycle with an acceleration of order $\omega_p c$, then the particle energy would be of order $E_0 \sim m_p c^2$, where $m_p = (\hbar c/G)^{1/2} \approx 10^{19}$ GeV/c^2 is the Planck mass, except that the particle would radiate its rest energy away in the order of one wave period. Thus, the conditions for sources of inertial radiation in the universe at frequencies above the order of ω_p seem to be unattainable.

From the same considerations that lead to the electromagnetic blackbody radiation formula, the average energy per inertial-radiation mode in thermal equilibrium is

$$\bar{\varepsilon}_\omega = \frac{\hbar\omega}{2} + \frac{\hbar\omega}{\exp(\hbar\omega/T) - 1}. \qquad (13.27)$$

where T is the temperature in energy units. That is, just as there is a cosmic microwave background (CMB) with a thermal blackbody spectrum at a temperature 2.7 K, there should also be a cosmic inertial noise background with a thermal blackbody spectrum. In a sense, this inertial noise background is much easier to detect than the CMB, and in fact its effects have been detected decades before the effects of the CMB were detected in radar static noise. If the inertial noise field accounts for the Schrödinger equation, then every measurement of the effects of the Schrödinger equation is an indirect measurement of the inertial noise background.

The inertial noise field can account not only for the Schrödinger equation, but also for the quantization of the inertial field. According to Eq. (6.12), a particle at rest in F will remain at rest in a zero-energy state until a specific force of $D = c^2 / R$ (in a spatially flat, static universe) is applied to it. Then it will be in the lowest energy state above zero. The wavenumber corresponding to that first excited state, according to Eq. (13.11), must be of order $k_1 \sim 1/R$. That is, the wavelength of the first excited state is of the order of the length scale of the observable universe, R. Then the frequency of the first excited state is of order $\omega_1 \sim c/R$, and its energy is of order

$$E_1 \sim \hbar\omega_1 \sim \hbar c / R \sim \hbar D / c. \qquad (13.28)$$

Lending support to these concepts and estimates, [58] found that a particle undergoing uniform acceleration a_0 satisfies a one-dimensional Schrödinger equation with quantized energy levels at integers n,

$$E_n = n\hbar a_0 / c, \qquad (13.29)$$

so that the first excited state is $E_1 = \hbar a_0 / c$, consistent with Eq. (13.28).

In summary, the internal motion of the universe produces a fluctuating inertial noise field everywhere. The stable ground-state energies of oscillators suggest that a dissipation mechanism damps their resonant response to inertial noise. If dipole inertial radiation is the dissipation mechanism, and if the spectral density of the inertial noise is about $l_p^2 \omega^3$ with a cutoff at the Planck frequency, then inertial noise can cause particles to undergo a Brownian motion described by the Schrödinger wave equation.

Chapter 14
Is the Inertial Field Mediated by a Massless Spin-1 Boson?

The inertial field shares significant features in common with the electromagnetic field. The features are so similar that they offer some reason to expect that, just as the electromagnetic field is quantized and mediated by photons, which are massless spin-1 bosons, the inertial field might also be quantized and mediated by massless spin-1 bosons.

Both the electromagnetic field and inertial field are vector fields capable of supporting dipole radiation. Both inertial fields and electromagnetic fields exert a vector force on particles. The equation of motion of a particle in either an inertial field or an electromagnetic field may be written, as in Eq. (6.1), as

$$du^\mu / d\tau + \Gamma^\mu_{\alpha\beta} u^\alpha u^\beta = f^\mu, \tag{14.1}$$

where $u^\mu = [cdt / d\tau, \mathbf{u}]$ is the velocity 4-vector of the particle, and f^μ is a 4-vector specific force. In the case of an inertial field, f^μ is the velocity-dependent inertial drag 4-vector derived in Eq. (6.8). In the case of an electromagnetic field, $f^\mu = (q/m)F^\mu_{\ \nu} u^\nu$ is the velocity-dependent electromagnetic 4-vector specific force acting on a particle of charge q and mass m.

The mode structure of the electromagnetic and inertial noise fields is the same, and as shown in Eq. (13.16), the mode density per unit angular frequency of the inertial noise field is the same as it is for the electromagnetic noise field.

Just as there is currently no evidence to suggest that the electromagnetic field propagates slower in vacuum than the limiting speed of the universe, there is currently no evidence to suggest that gravitational or inertial fields propagate more slowly than c. That is, there is currently no evidence to suggest that a particle that mediates the inertial field should not be massless.

As shown in Ch. 12, when a gravitationally unbound dipole or quadrupole is accelerated, it radiates power to the radiation zone. The radiated power communicates the new position and state of motion of the quadrupole to all the mass on the future light cone. The radiated power also provides the energy to the mass on the future light cone to adjust its position and velocity in response to the new position and velocity vectors of the quadrupole. In terms of a field description, the radiated power is transported by a perturbation of the inertial field with amplitude that falls with range r as r^{-1}. In terms of a particle description, the radiated power is transported by particles with energy flux and fluence that fall with range r as r^{-2}.

As shown in Ch. 4, the energy of an electromagnetic pulse propagating, for example, from a supernova is dissipated as the pulse does work against the inertial field of the mass in the universe. Energy dissipates from the electromagnetic pulse by attenuation through a process described in Ch. 4 that leads to a loss of fractional energy, rather than of energy itself, per unit distance. As shown in Ch. 5, the fractional energy dissipation by work done against the inertial field of the universe over a cosmologically significant distance x, as seen by a 'comoving' observer, is

$$\Delta E(x) / E_0 \approx \exp(-x/R) - 1,$$ (14.2)

as shown in Fig. 14.1, where $R = 13.8 \pm 1.2$ billion light years is the radius of the event horizon in a nearly flat, static universe, from Ch. 4.

After propagating a distance

$$x_{1/2} = R \ln 2 = 0.69 R,$$ (14.3)

half of the photons have been attenuated from the electromagnetic pulse from doing work against the inertial field of the universe. The half-life of photons for decay into particles mediating the inertial field in a nearly flat, static universe is

$$T_{1/2} = x_{1/2} / c = 9.6 \pm 0.8 \text{ billion years.}$$ (14.4)

As the photons decay, the particles mediating the inertial field provide the energy to the mass on the future light cone of the electromagnetic pulse to adjust the position and velocity of that mass in response to the new characteristics of the electromagnetic pulse.

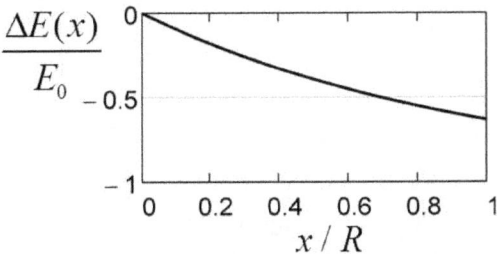

Fig. 14.1. Fractional energy loss from electromagnetic pulse *vs.* normalized range due to decay of photons into particles mediating the inertial field in nearly flat, static universe.

Chapter 15
Key Results

New exact solutions of Einstein's equation reveal that the basis for an inertial field is already embodied within classical general relativity. The inertial field in a *static, homogeneous universe* accounts for observations of anomalies that have been made at physical scales from our solar system to the cosmos. The inertial field can also account for physical phenomena at laboratory and atomic scales, such as Newton's third law and the effects of the Schrödinger wave equation. This section reviews the key results of this book.

An inertial field can account for these anomalies and physical phenomena whether or not our universe is expanding. That is, our universe does not need to be expanding in order for an inertial field to account for these phenomena. To the contrary, the model of an expanding universe raises issues concerning the cosmological principle and the Copernican principle, as discussed in Appendix F, that are not issues for an inertial field in a static universe, one that is not expanding at all. Moreover, the model of an expanding universe must posit 'new physics', such as dark energy and dark matter, with adjustable parameters. The model of a static universe with an inertial field emerges naturally from Einstein's equation with no 'new physics' and requires no adjustable parameters to account for these phenomena.

A static universe has a preferred reference frame F, the frame at rest with respect to all the mass. An inertial field in a static, homogeneous universe endows all of space uniformly with certain properties. These properties of an inertial field are perhaps best defined by their effects on the motion of mass with respect to the rest frame F. The inertial field acts on a mass moving through F with a constant force opposing the motion, an inertial drag force. The inertial field reacts instantaneously to a mass accelerating in F with a specific force that is, at least approximately, equal and opposite to the acceleration. This reaction force could be the reaction force of Newton's third law. Inertial radiation then transmits to the mass of the universe at the speed of light a net force equal and opposite to the reaction force.

The inertial field makes clocks seem to run slower the farther they are from us. The new solutions of Einstein's equation tell us this must be so even in a universe that is not expanding at all and no matter how much or how little curvature (intrinsic distortion of spacetime) the universe has. A Big Bang model of an expanding universe is not needed to account for the red shift of distant astronomical objects. The time dilation caused by the inertial field even in a static universe can account fully for distant objects appearing to move away from us.

Whether our universe is expanding or not, this inertial time dilation causes clocks to appear to stop completely at some finite range R. In an infinite homogeneous universe, one that is about the same everywhere, no observer anywhere can get information directly from beyond a range R in any direction. This maximum radius of our universe appears to be $R = 13.8 \pm 1.2$ billion light years. In this sense, our universe is like an inside-out black hole. Instead of being outside a black hole and unable to get information directly from the inside, we are inside our part of the universe and unable to get information directly from the outside. (We may, however, be able to get information from the outside relayed to us indirectly.)

When the universe is modeled as a collection of discrete masses, instead of a uniform continuous medium like a gas, the distinction blurs between the inertial field of the universe and the mass within the universe. The distribution of mass gives rise to the inertial field, and the inertial field endows matter with inertial mass. This ambiguity is perhaps what Einstein was expressing by "...the division into matter and field is ... something artificial and not clearly defined [8]."

Also ambiguous is where the gravitational field ends and the inertial field begins. The solution of Einstein's equation in a universe of discrete masses shows that the curvature of space and time at every point depends on the disposition of all the mass lying in the past of that spacetime point. That is, the very metric of spacetime, which is the framework of virtual clocks and measuring sticks in the universe, is completely determined by the disposition of mass in the past of each spacetime point. The components of the spacetime metric literally comprise descriptions of the disposition of mass throughout the universe.

Near a mass, the inertial field takes on the appearance of the Newtonian gravitational field of that mass. Alternatively, the gravitational field of a mass can be viewed as the contribution of that mass to the inertial field of the universe. This manifestation of the inertial field might be regarded as the embodiment of Einstein's hope that "we could regard matter as the regions in space where the field is extremely strong [8]."

Although the macroscopic effects of the inertial field are slight enough to have passed as anomalies or modifications of Newtonian gravity at very weak gravitational field strengths, the cumulative effects over cosmological distances are substantial. The cumulative effects of the inertial field can make a static universe not only appear to expand, but can make the speed of expansion appear to accelerate at cosmological distances, completely in accord with observations.

As light propagates across the universe, the inertial field affects its power and energy in three ways: (i) Inertial time dilation causes the frequency of light arriving at our sensors to be redshifted and its energy reduced; (ii) a pulse of light arriving at our sensors is stretched and its power reduced, also caused by the apparent slowing of clocks at a distance; and (iii) inertial drag causes a decay of energy and power from light with a half-life of about 10 billion years.

These three effects of the inertial field on light propagating across the universe completely account for measurements of the power and energy measured from Type Ia supernovas. These supernova measurements have been used, with adjustable parameters, to suggest that an expansion of our universe is accelerating. But these three effects of the inertial field alone account for the supernova measurements in a *static* universe, and they do so with *no adjustable parameters*. Moreover, unlike the model of an accelerating expansion, the model of an inertial field in a static universe does not violate the Copernican principle, the principle that we do not occupy a special, unique, or privileged observation point in the universe. Tests over the next decade or so might be able to distinguish whether the apparent cosmic acceleration is real or not.

An exact equation of motion was derived for particles in an inertial field. In one year, the inertial drag force decelerates particles moving much slower than the speed of light by about 2 cm/s.

The Pioneer anomaly is an apparent acceleration towards us of the two Pioneer spacecraft, Pioneer 10 and 11, that were launched over 40 years ago to depart our solar system. This acceleration, over 10 billion times weaker than the acceleration of gravity on Earth, is mostly constant, but with a small decaying component. Inertial time dilation accounts for the constant part. The smaller decaying part may be due to unrelated thermal effects. Inertial drag has negligible apparent effect on the spacecraft because inertial drag acts with about the same deceleration on the Earth and the Sun and the spacecraft. Neither inertial drag nor inertial time dilation affects the rate of precession of planetary orbits significantly, which is consistent with the very precise measurements of planetary precession rates.

Although inertial drag is very weak, in the outer regions of spiral galaxies, gravity is even weaker. In these outer regions, inertial drag dominates gravity and centrifugal forces and orbital velocities of stars around their galaxies. Although spiral galaxies appear to be rotating, for the most part inertial drag has stopped the outer regions from orbiting after just a fraction of an orbit. Once the outer regions stop rotating, they are supported from collapsing towards the center of the galaxy by the thermal pressure of the galaxy, rather than by centrifugal forces. Chapter 9 shows that pressure-supported galaxies have features that are attributed to dark matter or modified Newtonian dynamics (MOND), like flat rotation curves, mass proportional to the fourth power of (thermal) velocity, and a limit on mass per unit area of the galaxy.

Gravitationally-bound, pressure-supported systems, like plasma clouds, galaxies, and clusters, that are hot are slowed less by inertial drag than cooler systems. Inertial drag acts more to cool down hot systems than to slow their centers of mass. This phenomenon could explain the 'bow-shock' shape of the hot plasma associated with the Bullet cluster. The 'bow-shock' shape could be the result of the hot inner core of the plasma moving ahead of the cooler outer regions because it is decelerated less by inertial drag.

Acceleration of mass in every macroscopic gravitationally-*unbound* system produces inertial radiation. Unlike classical gravitational radiation, inertial radiation can cause isolated particles to accelerate. Although not calculated here, measurements suggest that gravitationally-*bound* systems, like binary pulsars, do not produce significant inertial radiation.

Only accelerating sources can produce inertial radiation. An approximate calculation shows that the force applied to accelerate a source of inertial radiation is in the same direction and about the same magnitude as the force the radiation applies on all the mass in the universe in the future. This approximate equivalence suggests that the reaction force of Newton's third law, equal and opposite to the applied force, is imposed by the inertial field instantaneously and is transmitted by the inertial radiation to the mass of the universe in the future.

Fluctuations of the inertial field of the universe, or 'inertial noise', can account classically for all the quantum mechanical effects described by the time-dependent Schrödinger wave equation. In equilibrium with the inertial noise field, particles absorb energy from fluctuations of the inertial field and radiate energy by dipole inertial radiation.

The similarities of the cosmic microwave background of the universe with the presumed properties of the cosmic inertial noise background suggest that the inertial field might be mediated by particles just as the electromagnetic field is mediated by photons, and that the particles might be massless, spin-1 bosons, just as photons are.

This book has shown that an inertial field in an Einstein *static* universe can account for important astrophysical observations that are generally attributed to effects expected in an *expanding* universe. Not addressed here are the questions of how and when such an effectively infinite, static universe could have been created. Appendix F addresses the issue of which matters are appropriate subjects of scientific inquiry according to the falsifiability criterion of [5], particularly with respect to the origin of a static universe.

Appendix A
Exact Inertial Field of Static, Homogeneous Universe

The most general spacetime interval in isotropic Cartesian coordinates, x, y, z, for a static, homogeneous spacetime is

$$c^2 d\tau^2 = e^{P(\mathbf{x})} c^2 dt^2 - e^{Q(\mathbf{x})} (dx^2 + dy^2 + dz^2),\qquad (A.1)$$

where $\mathbf{x} - \mathbf{x}_0$ is the displacement 3-vector from any origin, $\mathbf{x}_0$. That is, the only nonvanishing components of the metric tensor are the diagonal components,

$$g_{00} = e^P,\ g_{11} = g_{22} = g_{33} = -e^Q.\qquad (A.2)$$

The Christoffel symbols are

$$\Gamma^\alpha{}_{\mu\nu} = \left(g^{\alpha\beta}/2\right)\left(g_{\beta\mu,\nu} + g_{\nu\beta,\mu} - g_{\mu\nu,\beta}\right),\qquad (A.3)$$

where a comma denotes partial differentiation, as for example, $g_{\mu\nu,\beta} \equiv \partial g_{\mu\nu}/\partial x^\beta$. With this metric, 30 Christoffel symbols do not vanish identically, ten for each of the components, x, y, z. The ten nonvanishing Christoffel symbols for the x component are

$$\Gamma^0{}_{01} = \Gamma^0{}_{10} = P_x/2,\quad \Gamma^1{}_{00} = e^{P-Q} P_x/2$$
$$\Gamma^1{}_{11} = -\Gamma^1{}_{22} = -\Gamma^1{}_{33} = Q_x/2 \qquad (A.4)$$
$$\Gamma^1{}_{12} = \Gamma^1{}_{21} = Q_y/2,\quad \Gamma^1{}_{13} = \Gamma^1{}_{31} = Q_z/2,$$

where a subscript coordinate $x, y,$ or z indicates partial differentiation with respect to that coordinate, as for example, $P_x \equiv \partial P/\partial x$. The Christoffel symbols for the y and z components are found from Eq. (A.4) by cyclically permuting the indices, 1, 2, 3, and coordinates, x, y, z. The Riemann curvature tensor,

$$R^\alpha{}_{\beta\mu\nu} = -\Gamma^\alpha{}_{\beta\mu,\nu} + \Gamma^\alpha{}_{\beta\nu,\mu} + \Gamma^\sigma{}_{\beta\nu}\Gamma^\alpha{}_{\sigma\mu} - \Gamma^\sigma{}_{\beta\mu}\Gamma^\alpha{}_{\sigma\nu},\qquad (A.5)$$

has the six independent components,

$$R^0_{101} = -P_{xx}/2 - P_x^2/4 + P_x Q_x/2 - (\nabla P \cdot \nabla Q)/4$$
$$R^0_{202} = -P_{yy}/2 - P_y^2/4 + P_y Q_y/2 - (\nabla P \cdot \nabla Q)/4$$
$$R^0_{303} = -P_{zz}/2 - P_z^2/4 + P_z Q_z/2 - (\nabla P \cdot \nabla Q)/4$$
$$R^1_{212} = Q_{zz}/2 - Q_z^2/4 - (\nabla^2 Q)/2$$
$$R^1_{313} = Q_{yy}/2 - Q_y^2/4 - (\nabla^2 Q)/2$$
$$R^2_{323} = Q_{xx}/2 - Q_x^2/4 - (\nabla^2 Q)/2 \ .$$

(A.6)

The other components needed for the contraction of the Riemann tensor to the Ricci tensor, $R_{\beta\mu} = R^\alpha{}_{\beta\mu\alpha}$, are found from the identities $R_{\alpha\beta\mu\nu} = -R_{\beta\alpha\mu\nu}$ and $R_{\alpha\beta\mu\nu} = -R_{\alpha\beta\nu\mu}$. Then the only components of the Ricci tensor that do not vanish identically are the diagonal components,

$$R^0{}_0 = -e^{-Q}\left[\nabla^2 P/2 + (\nabla P)^2/4 + (\nabla P \cdot \nabla Q)/4\right]$$

(A.7a)

$$R^1{}_1 = -e^{-Q}\{(P_{xx} + Q_{xx})/2 + P_x Q_x/4$$
$$+[P_x^2 - Q_x^2 + (\nabla Q)^2]/4 + \nabla^2 Q/2\}$$

(A.7b)

$$R^2{}_2 = -e^{-Q}\{(P_{yy} + Q_{yy})/2 + P_y Q_y/4$$
$$+\left[P_y^2 - Q_y^2 + (\nabla Q)^2\right]/4 + \nabla^2 Q/2\}$$

(A.7c)

$$R^3{}_3 = -e^{-Q}\{(P_{zz} + Q_{zz})/2 + P_z Q_z/4$$
$$+\left[P_z^2 - Q_z^2 + (\nabla Q)^2\right]/4 + \nabla^2 Q/2\}^.$$

(A.7d)

The contraction of the Ricci tensor, $R^\sigma{}_\sigma$, the curvature scalar, is

$$C = -e^{-Q}\left[\nabla^2 P + (\nabla P)^2/2 + (\nabla P \cdot \nabla Q)/2 + 2\nabla^2 Q + (\nabla Q)^2/2\right].$$

(A.8)

Combining Eqs. (A.7a) and (A.8) eliminates P, as

$$C/2 - R^0{}_0 = -e^{-Q}\left[\nabla^2 Q + (\nabla Q)^2/4\right].$$

(A.9)

A scale factor k is defined as $k \equiv 2(1 - 2R^0{}_0/C)$. Then Eqs. (A.9) and (A.7a) become

$$4\nabla^2 Q + (\nabla Q)^2 = -kCe^Q \tag{A.10a}$$

$$2\nabla^2 P + (\nabla P)^2 + (\nabla P \cdot \nabla Q) = -(2-k)Ce^Q . \tag{A.10b}$$

Einstein's equation is

$$R^\mu_{\ \nu} = \left(C/2 - \Lambda/c^2\right)\delta^\mu_{\ \nu} - \kappa T^\mu_{\ \nu} , \tag{A.11}$$

where Λ is the cosmological constant, $\delta^\mu_{\ \nu}$ is the four-dimensional Kronecker delta, $T^\mu_{\ \nu}$ is the energy-momentum tensor, and $\kappa \equiv 8\pi G/c^4$. Symmetry requires $R^1_{\ 1} = R^2_{\ 2} = R^3_{\ 3}$ and $T^1_{\ 1} = T^2_{\ 2} = T^3_{\ 3}$. In a static spacetime, Einstein's equation requires $(k-1)C = \kappa(3T^0_{\ 0} - T^1_{\ 1} - T^2_{\ 2} - T^3_{\ 3})$, and both sides of this equation transform as a scalar only if $T^1_{\ 1} = T^2_{\ 2} = T^3_{\ 3} = 0$, so that $T^\sigma_{\ \sigma} = T^0_{\ 0}$. Then in terms of the scale factor k, the $T^0_{\ 0}$ component is given by the 'scaled curvature', $kC/4 = \kappa T^0_{\ 0} + \Lambda/c^2$. Einstein's equation gives the following relations for the scaled curvature, $kC/4$,

$$kC/4 = \kappa T^0_{\ 0} + \Lambda/c^2 = [3k/(4-k)]\Lambda/c^2 , \tag{A.12}$$

and for $\kappa T^0_{\ 0}$,

$$\kappa T^0_{\ 0} = (k-1)C/3 = [(k-1)/(1-k/4)]\Lambda/c^2 . \tag{A.13}$$

The first set of exact solutions considered in this appendix is the field of a static spacetime that possesses spherical symmetry, as that about a central mass. A solution of Eq. (A.10a) is

$$e^Q = \alpha_0(\alpha_0 + kCr^2/48)^{-2} , \tag{A.14}$$

where $\mathbf{r} = \mathbf{x} - \mathbf{x}_0$ is the radial displacement 3-vector from the origin, and α_0 is a constant. The condition $e^Q = 1$ at the origin requires $\alpha_0 = 1$. Different values of k correspond to different coordinate transformations of the metric, as exhibited in Eq. (A.14).

The solutions of Eqs. (A.10) are in some cases facilitated by a transformation to new variables, $p(\mathbf{x}) \equiv e^{P(\mathbf{x})/2}$ and $q(\mathbf{x}) \equiv e^{Q(\mathbf{x})/4}$. In terms of these new variables, Eqs. (A.10) become

$$\nabla^2 q = -(kC/16)q^5 \tag{A.15a}$$

$$q\nabla^2 p + 2\nabla p \cdot \nabla q = -(2-k)(C/4)pq^5 . \tag{A.15b}$$

This appendix considers three cases, outlined in Table A.1, corresponding to solutions of Eqs. (A.10) for distinct coordinate transformations.

Table A.1. Three cases of coordinate transformations (values of k) are considered.

Case	R^0_0	k	$kC/4$	κT^0_0	Static Universe
0	0	2	$3\Lambda/c^2$	$+C/3$	Friedmann-Lemaître
1	$C/4$	1	Λ/c^2	0	Empty-Lemaître
2	$C/2$	0	0	$-C/3$	Flat 3-Volume

Case 0. Friedmann-Lemaître Static Universe

If $R^0_0 = 0$, then $k = 2$, and a trivial solution of Eq. (A.10b) is $e^P = 1$, which leads to the Friedmann-Lemaître model of a static universe with $T^0_0 \neq 0$ and $\Lambda \neq 0$. More generally, however, for $R^0_0 = 0$ and $R^2 \equiv 24/C$, Eq. (A.10b) combined with Eq. (A.14) for $k = 2$ becomes

$$\frac{dP_r}{dr} + \frac{2P_r}{r(1+r^2/R^2)} + \frac{P_r^2}{2} = 0, \tag{A.16}$$

which is the first-order Bernoulli equation [7] in the variable $P_r \equiv dP/dr$. A general solution of Eq. (A.16) in these spherical coordinates is

$$e^{P/2} = \alpha_1 + (\alpha_2/r)(1 - r^2/R^2), \tag{A.17}$$

where α_1 and α_2 are integration constants. For $\alpha_2 = 0$ and $\alpha_1 = 1$, a solution for a static, homogeneous spacetime with $R^0_0 = 0$, the Friedmann-Lemaître model of a static universe, is shown in Fig. A.1(a).

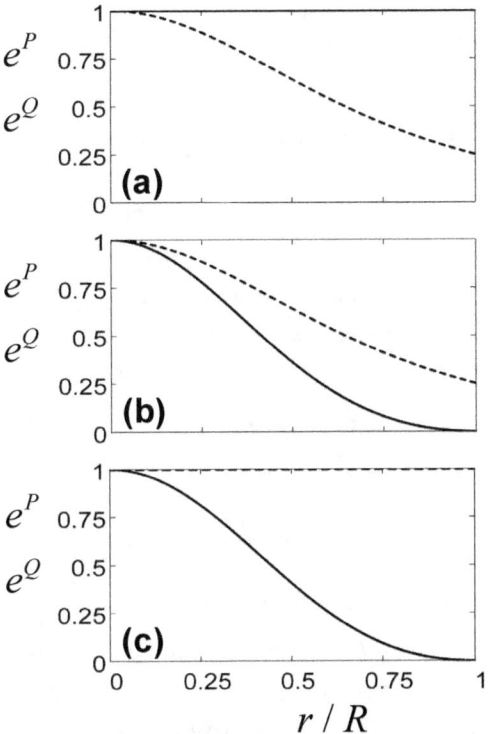

Fig. A.1. Exact solutions for static, homogeneous spacetime with $C > 0$ in spherical coordinates, e^P (solid) and e^Q (dashed) *vs.* r / R for: (a) $k = 2$; (b) $k = 1$; (c) $k = 0$.

Case 1. Empty-Lemaître Static Universe

If $R^0_0 = C / 4$, then $k = 1$, corresponding to an empty-Lemaître model of a static universe with $T^0_0 = 0$ and $\Lambda \neq 0$. For $R^2 \equiv 48 / C$, Eq. (A.10b) combined with Eq. (A.14) for $k = 1$ becomes

$$\frac{dP_r}{dr} + \frac{2P_r}{r(1 + r^2 / R^2)} + \frac{P_r^2}{2} = \frac{-24 / R^2}{(1 + r^2 / R^2)^2}, \qquad (A.18)$$

which is the first-order Riccati equation [7] in the variable $P_r \equiv dP / dr$. With the substitution, $p = e^{P/2}$, the Riccati equation becomes the second-order linear equation,

$$\frac{d^2 p}{dr^2} + \frac{2R^2}{r(R^2 + r^2)} \frac{dp}{dr} + \frac{12R^2}{(R^2 + r^2)^2} p = 0. \tag{A.19}$$

A general solution of Eq. (A.19) in these spherical coordinates is

$$e^{P/2} = \left(R^2 + r^2\right)^{-1} \left[\alpha_1\left(R^2 - r^2\right) + \left(\alpha_2 / r\right)\left(1 - 6r^2 / R^2 + r^4 / R^4\right)\right], \tag{A.20}$$

where α_1 and α_2 are integration constants. For $\alpha_1 = 1$ and $\alpha_2 = 0$, a solution for a static, homogeneous spacetime with $R^0_{\ 0} = C/4$,

$$e^P = \left(\frac{1 - r^2 / R^2}{1 + r^2 / R^2}\right)^2 = \tanh^2\left[\ln(r/R)\right], \tag{A.21}$$

is shown in Fig. A.1(b).

Case 2. Flat-3-Volume Static Universe

If $R^0_{\ 0} = C/2$, then $k = 0$, corresponding to zero scaled curvature, $\kappa T^0_{\ 0} + \Lambda / c^2 = 0$, which is not the same as zero curvature. For this case, since $C = -3\kappa T^0_{\ 0}$ and $\Lambda / c^2 = -\kappa T^0_{\ 0}$, a positive energy density, $T^0_{\ 0} > 0$, requires negative curvature, $C < 0$, and $\Lambda < 0$. In the following, exact solutions are given first for positive curvature and then for negative curvature.

For positive curvature, defining $R^2 \equiv 2\pi^2 / C$, Eq. (A.10b) combined with Eq. (A.14) for $k = 0$ becomes

$$\frac{dP_r}{dr} + \frac{2P_r}{r} + \frac{P_r^2}{2} = \frac{-2\pi^2}{R^2}, \tag{A.22}$$

which is again the first-order Riccati equation [7] in the variable $P_r \equiv dP / dr$. With the substitution, $p = e^{P/2}$, the Riccati equation becomes the second-order linear equation,

$$\frac{d^2 p}{dr^2} + \frac{2}{r}\frac{dp}{dr} + \frac{C}{2} p = 0. \tag{A.23}$$

A general solution of Eq. (A.23) in these spherical coordinates for $C > 0$ is

$$e^{P/2} = \alpha_1(R/\pi r)\sin(\pi r / R) + \alpha_2(R/\pi r)\cos(\pi r / R), \tag{A.24}$$

where α_1 and α_2 are integration constants. For $\alpha_1 = 1$ and $\alpha_2 = 0$, a solution for a static, homogeneous spacetime with $R^0_{\ 0} = C/2 > 0$,

$$e^P = (R/\pi r)^2 \sin^2(\pi r / R),\qquad\qquad\qquad\qquad (A.25)$$

is shown in Fig. A.1(c).

The same case, $R^0_{\ 0} = C/2$ and $k = 0$, but with negative curvature, still corresponds to zero scaled curvature, $\kappa T^0_{\ 0} + \Lambda/c^2 = 0$, but positive energy density, $T^0_{\ 0} > 0$. Equation (A.10b) combined with Eq. (A.14) for $k = 0$ transforms to Eq. (A.23), as before, but a general solution of Eq. (A.23) in these spherical coordinates with $C < 0$ is

$$e^{P/2} = (\alpha_1 L / r)\sinh(r / L) + (\alpha_2 L / r)\cosh(r / L),\qquad\qquad (A.26)$$

where α_1 and α_2 are integration constants, and $L \equiv (-2/C)^{1/2}$ is a constant length. For $\alpha_1 = 1$ and $\alpha_2 = 0$, a solution for a static, homogeneous spacetime with $R^0_{\ 0} = C/2 < 0$ is

$$e^P = (L/r)^2 \sinh^2(r / L).\qquad\qquad\qquad\qquad (A.27)$$

Each of the general solutions $e^{P/2}$ in Eqs. (A.17), (A.20), (A.24), and (A.26) has a term that diverges at the origin as $1/r$. In a homogeneous universe, there should be no singularities in the metric. The apparent singularities in these solutions arise from the *de facto* spherical coordinate system, which has a $1/r$ term in the Laplacian operator, $\nabla^2 = d^2/dr^2 + (2/r)d/dr$. The apparent singularities are an artifact of imposing spherical coordinates on a homogeneous universe that has no natural origin.

Spherical coordinates are useful for dealing with a central mass. For example, the vacuum Einstein equations, $\nabla^2 q = 0$ and $\nabla^2(pq) = 0$, have the solution, $q = 1 + r_s / r$ and $pq = 1 - r_s / r$, where $r_s \equiv Gm_0 / 2c^2$ is the Schwarzschild radius of a central mass m_0 in isotropic spherical coordinates (see Ch. 3). But a spherically symmetric spacetime, such as that surrounding a central mass, has only two degrees of freedom, both rotational. A homogeneous universe has three translational degrees of freedom. Needlessly constraining the third degree of freedom with spherical coordinates introduces the $1/r$ term in the solutions.

A general solution of Eq. (A.15b) without singularities can be found in Cartesian coordinates. The solution is valid along every ray from the origin. The

g_{00} component of the metric of a static, homogeneous spacetime for an observer at the origin can then be defined through the solution of Eq. (A.15b) along the set of all rays through the origin. With the choice of coordinates corresponding to Case 2, such that $k = 0$ and $q = 1$, Eq. (A.15b) becomes

$$\nabla^2 p + (C/2)p = 0 .$$ (A.28)

A singularity-free solution of Eq. (A.28) along any ray, such as the $+x$ axis for example, for $C < 0$ is

$$p = \alpha_1 \sinh[(-C/2)^{1/2} x] + \alpha_2 \cosh[(-C/2)^{1/2} x] ,$$ (A.29)

and for $C > 0$ is

$$p = \alpha_1 \sin[(C/2)^{1/2} x] + \alpha_2 \cos[(C/2)^{1/2} x] .$$ (A.30)

If the observer is at the origin, where $g_{00} = 1$, and the other boundary condition is $g_{00} = 0$ at $x = R$, and if the energy-conservation condition $dp/dx \leq 0$ from $x = 0$ to $x = R$ is satisfied, then the exact diagonal metric of the inertial field of a static, homogeneous universe in these coordinates may be expressed along any ray from the origin as

$$g_{00} = \frac{\sinh^2[C_0(1 - x/R)]}{\sinh^2 C_0} , \quad g_{11} = g_{22} = g_{33} = -1 ,$$ (A.31)

where $C_0 \equiv (|C|R^2/2)^{1/2}$ is a dimensionless curvature parameter. For $C > 0$, the hyperbolic sine functions in Eq. (A.31) may be replaced by sine functions of the same arguments.

The g_{00} component in Eq. (A.31) is a solution of the Poisson equation, Eq. (A.28), in the same sense that $\alpha_1 - \alpha_2 x$ is a solution of Laplace's equation. The solution is spherically symmetric even though the x coordinate is Cartesian and not curvilinear, because the x axis can be rotated to lie along any ray from the origin. When calculating the motion of particles, as in Ch. 6, proper account must be taken of the fact that x is merely a Cartesian measure of distance along any ray from the origin, as was illustrated in Fig. 2.1.

Appendix B
Effects of Weak Inertial Drag on Orbital Motion

This appendix calculates the effects of weak inertial drag on a particle moving in a gravitationally bound orbit about a mass m_0 in an otherwise static, homogeneous universe. The calculations are done for two cases: (1) An orbit about a mass m_0 at rest in the rest frame F of the static universe; and (2) an orbit at an orbital speed much slower than the drift velocity of the mass m_0 in F.

Case 1. An Orbit about a Mass at Rest in F

In isotropic spherical coordinates, r, θ, ϕ, the exact spacetime interval of a mass m_0 at the origin of a static, homogeneous universe is

$$c^2 d\tau^2 = g_{00} c^2 dt^2 + g_{11} \left(dr^2 + r^2 d\theta^2 + r^2 \sin^2 \theta d\phi^2 \right), \tag{B.1}$$

where the exact diagonal spacetime metric is given by Eq. (3.4) as

$$g_{00} = \left(1 + \frac{r_S}{r} \right)^{-2} \left(\frac{\sinh[C_0 (1 - x / R)]}{\sinh C_0} - \frac{r_S}{r} \right)^2, \tag{B.2}$$

$$g_{11} = g_{22} = g_{33} = -\left(1 + r_S / r \right)^4,$$

and where $r_S \equiv Gm_0 / 2c^2$ is the Schwarzschild radius in isotropic coordinates, $C_0 \equiv (|C| R^2 / 2)^{1/2}$ is the curvature parameter defined at Eq. (2.12), and the Cartesian coordinate x is distinguished from the radial spherical coordinate r as a measure of distance along any ray from the observer, as in Eq. (A.31).

From Ch. 6, the exact equation of motion of the orbiting particle in a static inertial field is

$$\dot{u}^\mu + \Gamma^\mu_{\alpha\beta} u^\alpha u^\beta = d^\mu, \tag{B.3}$$

where $u^\mu = [c\dot{t}, \mathbf{u}]$ is the velocity 4-vector of the particle, and an overdot indicates differentiation with respect to proper time τ, that is, $d / d\tau$. With the metric and isotropic coordinates of Eq. (B.2), the inertial drag 4-vector is given by Eq. (6.8) as

$$d^\mu = -D\left[W\dot{s}/c, W^{-1}\dot{i}\hat{s}\right],$$
(B.4)

where $\hat{s}$ is a unit 3-vector in the direction of instantaneous particle velocity $\dot{s} \equiv \mathbf{u}$, s is a coordinate measure of the total distance traveled by the particle in F, and

$$W(r,x) \equiv \left(1+\frac{r_S}{r}\right)^3 \left(\frac{\sinh[C_0(1-x/R)]}{\sinh C_0} - \frac{r_S}{r}\right)^{-1}.$$
(B.5)

To isolate the effects of inertial drag on orbits, consider a nonrelativistic particle orbiting in the weak Newtonian gravitational field of a mass m_0. That is, terms of higher order than $\dot{s}^2/c^2$ and Gm_0/rc^2 are neglected. In this approximation, there is no difference between the specific momentum of the orbiting particle, $\mathbf{u} = \dot{s}$, and its coordinate velocity, $\mathbf{v} = ds/dt$. Moreover, let the range x of the observer to the orbiting particle satisfy $r \ll x \ll R$, so that inertial time dilation and its variation over the orbit is negligible to the observer. In this modified-Newtonian approximation that neglects relativistic effects but allows for inertial drag, the energy equation from Eqs. (B.1) and (B.2) becomes

$$c^2 = (c^2 - 2Gm_0/r)\dot{t}^2 - [\dot{r}^2 + r^2\dot{\theta}^2 + (r\sin\theta)^2\dot{\phi}^2].$$
(B.6)

From Eq. (6.5) and Eqs. (B.3) to (B.6), the time component of the equation of motion in this same modified-Newtonian approximation is

$$c^2\ddot{t} + 2(Gm_0/r^2)\dot{r}\dot{t} = -D\dot{s}.$$
(B.7)

From Eq. (6.17), the specific energy of the particle, including the rest mass energy, in this modified-Newtonian approximation is

$$\gamma c^2 = c^2 - Gm_0/r + \dot{s}^2/2,$$
(B.8)

and from Eq. (6.16), the rate of loss of specific energy is

$$\dot{\gamma}c^2 = -D\dot{s}.$$
(B.9)

Integrating Eq. (B.9) and combining with Eq. (B.8) gives the Newtonian orbital energy equation, modified by dissipative inertial drag, as

$$E(s) = -Gm_0 / r + \dot{s}^2 / 2 = E_0 - Ds \,, \tag{B.10}$$

where $E_0 \equiv -Gm_0 / r_0 + \dot{s}_0^2 / 2 < 0$ is the initial (at $s = 0$) Newtonian specific energy $E(s)$ at the initial particle speed $\dot{s}_0$ and radius r_0.

From Eq. (6.6), the equation of motion in this modified-Newtonian approximation is

$$\ddot{\mathbf{s}} + (Gm_0 / r^2)\hat{\mathbf{r}} = -D\hat{\mathbf{s}} \,. \tag{B.11}$$

In terms of cylindrical radial and azimuthal coordinates, r and ϕ, and coordinate speed $v = (\dot{r}^2 + r^2\dot{\phi}^2)^{1/2}$, and particle direction $\hat{\mathbf{s}} = (dr / ds)\hat{\mathbf{r}} + (rd\phi / ds)\hat{\boldsymbol{\phi}}$, the equation of motion is

$$\begin{aligned}
\ddot{r} - r\dot{\phi}^2 + Gm_0 / r^2 &= -D(dr / ds) \\
r\ddot{\phi} + 2\dot{r}\dot{\phi} &= -D(rd\phi / ds)
\end{aligned} \tag{B.12}$$

In terms of the specific angular momentum of the orbiting particle, $l = r^2\dot{\phi}$, Eqs. (B.12) become

$$\ddot{r} - l^2 / r^3 + Gm_0 / r^2 = -D(dr / ds), \tag{B.13}$$

$$\dot{l} = -D(r^2 d\phi / ds) \,. \tag{B.14}$$

If the inertial drag constant D is comparable to or greater than Gm_0 / r^2 or v^2 / r, then inertial drag will have a substantial effect on the orbital dynamics, as will be shown in Ch. 9. For example, Ch. 9 will show that inertial drag can account for the flattening of rotation curves at ranges beyond the gravitational reach of the mass in disk galaxies, where D exceeds Gm_0 / r^2. On the other hand, if $D \ll Gm_0 / r^2$ and $D \ll v^2 / r$, then the effects of weak inertial drag on orbital dynamics can be calculated by a simple perturbation analysis, as in the following.

The perturbation analysis starts with the zeroth-order solution of Eqs. (B.10) and (B.14) for $D = 0$. In the absence of inertial drag, the particle orbit is an ellipse,

$$r = a_0(1-\varepsilon_0^2)/(1+\varepsilon_0\cos\phi),\tag{B.15}$$

where a_0 is the constant semi-major radius and ε_0 is the constant eccentricity of the ellipse. Figure B.1 shows an elliptical orbit corresponding to Eq. (B.15) for $\varepsilon_0 = 0.6$. The arc length of an ellipse is $s(\phi) = \int_0^\phi (ds/d\phi')\,d\phi'$, where

$$ds/d\phi = a_0(1-\varepsilon_0^2)(1+\varepsilon_0^2+2\varepsilon_0\cos\phi)^{1/2}/(1+\varepsilon_0\cos\phi)^2.\tag{B.16}$$

The perimeter p_ε of an ellipse, shown in Fig. B.2, is in the range $2\pi a_0 \geq p_\varepsilon \geq 4a_0$ for $0 \leq \varepsilon \leq 1$. The constant energy of the particle for $D = 0$ is

$$E_0 = -Gm_0/2a_0,\tag{B.17}$$

independent of eccentricity and angular momentum. Evaluating the energy at either apsis of the orbit gives the constant specific angular momentum l_0 as

$$l_0^2 = 2a_0^2(1-\varepsilon_0^2)|E_0| = Gm_0a_0(1-\varepsilon_0^2).\tag{B.18}$$

Combining Eq. (B.15) and (B.18) gives the azimuthal speed $r\dot\phi$ as

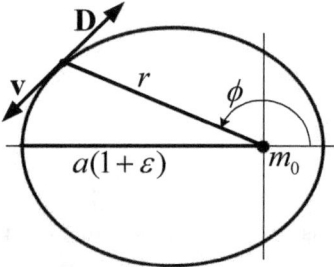

Fig. B.1. Elliptical orbit about mass m_0 at rest in F.

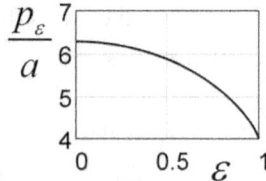

Fig. B.2. Perimeter of ellipse normalized to semi-major radius *vs.* eccentricity.

$$(r\dot{\phi})^2 = (Gm_0/r)(1+\varepsilon_0 \cos\phi),$$ (B.19)

the radial speed $\dot{r}$ as

$$\dot{r}^2 = (Gm_0/r)(\varepsilon_0 \sin\phi)^2/(1+\varepsilon_0 \cos\phi),$$ (B.20)

and the specific kinetic energy as

$$v^2/2 = |E_0|(1+2\varepsilon_0 \cos\phi+\varepsilon_0^2)/(1-\varepsilon_0^2).$$ (B.21)

From this zeroth-order analysis and Eq. (B.10), the first-order perturbation analysis in D gives the changes of energy, angular momentum, semi-major radius, and eccentricity over an orbital path length Δs for $D\Delta s \ll Gm_0/r$ and $D\Delta s \ll v^2$. From Eq. (B.10), the decay of orbital energy to first order in Δs is

$$\Delta E/|E_0| = -D\Delta s/|E_0|.$$ (B.22)

From Eq. (B.17), the decay of semi-major radius with Δs to first order is

$$\Delta a/a_0 = -D\Delta s/|E_0|.$$ (B.23)

From Eq. (B.14), the fractional rate of decay of angular momentum,

$$\dot{l}/l_0 = -D/v,$$ (B.24)

depends on eccentricity. From Eqs. (B.14) and (B.16), the decay of angular momentum with Δs to first order in $D\Delta s/|E_0|$ is

$$\frac{\Delta l}{l_0} = \frac{-D\Delta s}{2|E_0|}\left(\frac{1-\varepsilon_0^2}{1+2\varepsilon_0 \cos\phi+\varepsilon_0^2}\right).$$ (B.25)

The fractional decay of angular momentum *per orbit*,

$$\frac{\Delta l_1}{l_0} = \frac{-Dp_\varepsilon(\varepsilon_0)}{2|E_0|}\int_0^{2\pi}\frac{d\phi}{2\pi}\frac{r^2(\phi)/a_0^2}{(1+2\varepsilon_0 \cos\phi+\varepsilon_0^2)^{1/2}},$$ (B.26)

shown in Fig. B.3, is proportional to the perimeter $p_\varepsilon(\varepsilon_0)$, that is, to Δs for one orbit, and also depends on eccentricity. For small ε_0, a good approximation for

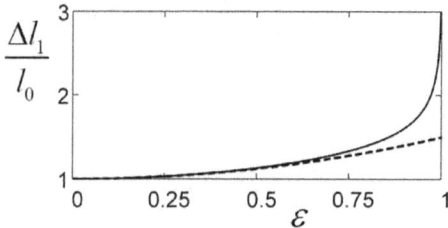

Fig. B.3. Fractional decay of angular momentum per orbit normalized to $-Dp_\varepsilon / 2|E_0|$ (solid), and $1 + \varepsilon^2 / 2$ (dashed) *vs.* eccentricity.

the decay of angular momentum with Δs to first order is

$$\Delta l / l_0 \approx -(D\Delta s / 2|E_0|)(1 + \varepsilon_0^2 / 2) . \tag{B.27}$$

From Eq. (B.18), the decay of angular momentum is related to the decay of semi-major radius and eccentricity by

$$\frac{\Delta l}{l_0} = \frac{\Delta a}{2a_0} - \frac{\varepsilon_0 \Delta \varepsilon}{1 - \varepsilon_0^2} . \tag{B.28}$$

From Eqs. (B.23), (B.27), and (B.28), the growth of eccentricity with Δs, to first order and for small eccentricity, is

$$\Delta \varepsilon / \varepsilon_0 \approx +D\Delta s / 4|E_0| . \tag{B.29}$$

<u>Case 2.</u> An Orbit Drifting through F Much Faster than Orbital Speed

In Case 1, the central mass m_0 was at rest in the rest frame F of a static universe. In Case 2, m_0 is moving through F at a constant velocity $\mathbf{V}_0$ much faster than the orbital velocity of a particle about m_0. Let F_0 represent the frame of reference in which m_0 is at rest, and let $\mathbf{v}$ be the velocity of a particle orbiting about the stationary mass m_0 in F_0. With the same modified-Newtonian approximation used for Case 1, suppose $v \ll V_0 \ll c$. That is, this case calculates the orbital motion of a particle about a mass m_0 drifting through the rest frame F of a static universe at a velocity $\mathbf{V}_0$ much faster than the orbital velocity $\mathbf{v}$ in the rest frame of m_0, so that terms of order $D(v / V_0)^2$ may be neglected. In this case, the modified-Newtonian approximation means even terms of higher order than V_0^2 / c^2 and Gm_0 / rV_0^2 are neglected. Also, the inertial-drag constant D will be taken to satisfy $D \ll Gm_0 / r^2$ and $D \ll v^2 / r$, so that a first-order perturbation analysis in $D\Delta s / |E_0|$ is appropriate.

Within the modified-Newtonian approximation, the velocity of the orbiting particle in F, the rest frame of the universe, is given by a Galilean transformation as $\mathbf{V}_0 + \mathbf{v}(t)$. In the rest frame F_0 of the central mass m_0, the motion of the particle will be calculated in cylindrical coordinates, r, ϕ, z. Just as the zeroth-order solution of Case 1 was simplified by the choice of $\theta = \pi / 2$ (and $\dot{\theta} = 0$) to define the orbital plane, the zeroth-order solution of this case is simplified by the choice of $z = 0$ (and $\dot{z} = 0$) to define the orbital plane. The velocity of the particle in F_0 to zeroth order, therefore, is $\mathbf{v} = \dot{r}\hat{\mathbf{r}} + (r\dot{\phi})\hat{\boldsymbol{\phi}} + 0\hat{\mathbf{z}}$, where a cap denotes a unit vector, and r is now a cylindrical, not a spherical, radial coordinate. In F_0, the coordinate origin, $r = 0$ and $z = 0$, will be defined to be at the stationary mass m_0. Damped, driven small-amplitude oscillations of the orbital plane along the z axis are interesting, but are not analyzed here.

In F, the rest frame of the static universe, let the constant velocity of the mass m_0 be

$$\mathbf{V}_0 = V_0 [\,\hat{\mathbf{x}}\sin\theta_0 \cos\phi_0 \,,\; \hat{\mathbf{y}}\sin\theta_0 \sin\phi_0 \,,\; \hat{\mathbf{z}}\cos\theta_0 \,]. \tag{B.30}$$

That is, the direction of $\mathbf{V}_0$ in F is at a polar angle θ_0 with respect to the z axis, which is normal to the orbital plane of the particle, and is at an azimuthal angle ϕ_0 with respect to the periapsis of the particle orbit at $\phi = 0$, as shown in Fig. B.4.

From Eq. (6.6), the equation of motion in F_0 in this modified-Newtonian approximation is

$$d\mathbf{v} / dt + Gm_0 (\mathbf{r} + \mathbf{z}) / (r^2 + z^2)^{3/2} = \mathbf{f}_0(t), \tag{B.31}$$

where $\mathbf{r}$ and $\mathbf{z}$ are the position vectors of the orbiting particle in the orbital plane and normal to the plane, respectively, with respect to the stationary origin at m_0 in F_0, and $\mathbf{f}_0(t)$ is the specific inertial drag force acting on the orbiting particle in F_0.

In F, the specific inertial drag force acting on the orbiting particle, from Eq. (7.4), is $\mathbf{f} = -D\hat{\mathbf{s}}$, where $\hat{\mathbf{s}}$ is a unit vector in the direction of particle motion in F, not F_0. That is, in F the specific inertial drag force on the orbiting particle is

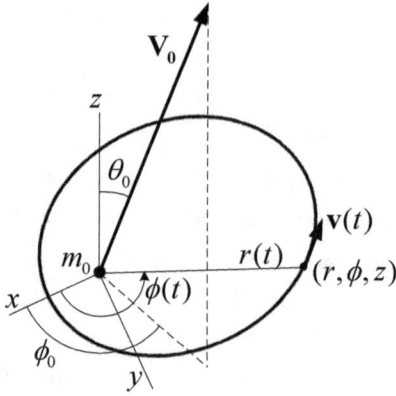

Fig. B.4. Elliptical orbit about mass m_0, which is at rest in F_0 and moving with constant velocity $\mathbf{V}_0$ in F.

$$\mathbf{f} = -D\left(\mathbf{V}_0 + \mathbf{v}(t)\right)/\left|\mathbf{V}_0 + \mathbf{v}(t)\right|,\tag{B.32}$$

and the specific inertial drag force on the mass m_0 in F is

$$\mathbf{f_m} = -D\hat{\mathbf{V}}_0,\tag{B.33}$$

where $\hat{\mathbf{V}}_0 = \mathbf{V}_0 / V_0$ is a unit vector in the direction of $\mathbf{V}_0$. In F_0, therefore, the specific inertial drag force on the orbiting particle transforms as

$$\mathbf{f}_0 = \mathbf{f} - \mathbf{f_m} = D\left[\hat{\mathbf{V}}_0 - \left(\mathbf{V}_0 + \mathbf{v}(t)\right)/\left|\mathbf{V}_0 + \mathbf{v}(t)\right|\right].\tag{B.34}$$

For $z = 0$ and neglecting terms of order $D(v/V_0)^2$, the equation of motion in F_0, Eq. (B.31), becomes

$$d\mathbf{v}/dt + Gm_0\mathbf{r}/r^3 = (D/V_0)\hat{\mathbf{V}}_0 \times [\hat{\mathbf{V}}_0 \times \mathbf{v}(t)].\tag{B.35}$$

The inertial drag force has constant magnitude in F, but in F_0 it is proportional to orbital speed. The direction of $\mathbf{V}_0$ with respect to the direction of the orbiting particle is

$$\hat{\mathbf{V}}_0 = \hat{\mathbf{r}}\sin\theta_0\cos(\phi-\phi_0) - \hat{\boldsymbol{\phi}}\sin\theta_0\sin(\phi-\phi_0) + \hat{\mathbf{z}}\cos\theta_0.\tag{B.36}$$

Combining Eqs. (B.35) and (B.36), the components of the equation of motion in terms of the specific angular momentum of the orbiting particle, $l = r^2\dot{\phi}$, become

$$\ddot{r} - l^2/r^3 + Gm_0/r^2 = -(D/V_0)[\dot{r} - v_0 \sin\theta_0 \cos(\phi - \phi_0)],\tag{B.37}$$

$$\dot{l} = -(D/V_0)[l_0 + rv_0 \sin\theta_0 \sin(\phi - \phi_0)],\tag{B.38}$$

where $v_0 \equiv \hat{\mathbf{V}}_0 \cdot \mathbf{v}(t)$. To zeroth order ($D = 0$), v_0 is a constant,

$$v_0 = -l_0 \sin\theta_0 / r(\phi_0),\tag{B.39}$$

where $r(\phi_0) = a_0(1 - \varepsilon_0^2)/(1 - \varepsilon_0 \cos\phi_0)$. The perturbation terms on the right-hand sides of Eqs. (B.37) and (B.38) can be evaluated by substituting Eq. (B.39) and the zeroth-order solutions for $r(\phi)$ and $\dot{r}(\phi)$,

$$r(\phi) = \frac{a_0(1 - \varepsilon_0^2)}{1 + \varepsilon_0 \cos\phi},\tag{B.40}$$

$$\dot{r}(\phi) = \frac{\varepsilon_0 l_0 \sin\phi}{a_0(1 - \varepsilon_0^2)}.\tag{B.41}$$

Regardless of the orientation of the orbital plane, as long as $v \ll V_0$, the inertial drag force on the particle in F_0 is smaller than it is in F by a factor at least of the order of v/V_0.

The zeroth-order equations (for $D = 0$) governing the motion of the orbiting particle in F_0 are the same as Eqs. (B.13) and (B.14) for Case 1, and the zeroth-order solutions are the same as Eqs. (B.15) – (B.21) for Case 1. Then the first-order perturbation analysis in D gives the small changes of energy, angular momentum, semi-major radius, and eccentricity over a time $t \ll V_0/D$. For a solar system moving through F with a speed $V_0 = 600$ km/s, this approximation is valid for times much shorter than 10 million years.

In Eq. (B.38), the second term on the right is periodic and causes no long-term change to angular momentum. But the first term on the right causes a linear decay with time as

$$l(t) = l_0(1 - Dt/V_0).\tag{B.42}$$

Similarly, the right-hand side of Eq. (B.37) is periodic and causes a much less significant long-term change to the orbital radius than the angular momentum term on the left. Ignoring the right-hand side then, in the new variable, $U \equiv 1/r$, Eq. (B.37) becomes

$$d^2U / d\phi^2 + U = Gm_0 / l^2(t), \tag{B.43}$$

which has the approximate solution,

$$r = a(t)[1 - \varepsilon^2(t)] / [1 + \varepsilon(t)\cos\phi], \tag{B.44}$$

where the semi-major radius and eccentricity decay as $a(t) = a_0(1 - 2Dt/V_0)$ and $\varepsilon(t) = \varepsilon_0(1 - Dt/V_0)$. And since $E(t) = -Gm_0 / 2a(t)$, the orbital energy decays as $E(t) = -(Gm_0 / 2a_0)(1 + 2Dt/V_0)$. This approximate solution, Eq. (B.44), was compared to the exact numerical solution of Eqs. (B.37) and (B.38) in Fig. 8.1.

Appendix C
Gravitational Field in a Disk Galaxy

This appendix calculates the gravitational field in the plane of a thin disk with an exponential density profile. A disk galaxy is modeled as a thin, cylindrically symmetric disk comprising many thin circular rings. Figure C.1 shows the configuration for calculating the field in cylindrical coordinates, (ρ, ϕ, z), of one thin ring of mass Δm and radius $\rho = b$. In the plane of this ring ($z = 0$), the radial Newtonian field of this ring at radius r is

$$\Delta g(r) = -G\Delta m \int_0^{2\pi} \left(\frac{r - b\cos\phi}{s^3} \right) \frac{d\phi}{2\pi}, \tag{C.1}$$

where $s \equiv (r^2 - 2br\cos\phi + b^2)^{1/2}$. Integrating Eq. (C.1) over r gives the potential of the ring in the plane of the ring,

$$\Delta\Phi(r) = -G\Delta m \int_0^{2\pi} \frac{d\phi}{2\pi s} = -\frac{G\Delta m}{(r+b)} \frac{2}{\pi} K(k) = -\frac{G\Delta m}{r} f(r), \tag{C.2}$$

where $K(k)$ is the complete elliptic integral of the first kind with modulus (also called 'parameter') $k \equiv 2(rb)^{1/2}/(r+b)$, and where

$$f(r) \equiv (2/\pi)K(k)/(1+b/r) \tag{C.3}$$

is a dimensionless 'enhancement factor' over what the potential of the ring would be if its mass Δm were concentrated in a point at the center. In the plane of a ring, the field is enhanced relative to the same mass in a spherical shell because the field components are concentrated in the plane.

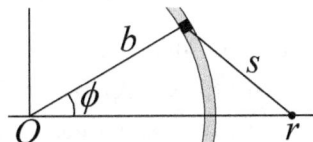

Fig. C.1. Configuration for calculating field at r of a thin ring at radius b.

Figure C.2 shows the enhancement factor f, the gravitational potential $\Delta\Phi$, and the gravitational field Δg in the plane of a ring. The field from a ring at $b < r$ is purely centripetal, and the field from a ring at $b > r$ is purely centrifugal. That is, a particle at r in the plane of the ring is attracted towards the closest point on the ring, whether the ring is inside or outside r. Figure C.2 shows that Newton's shell theorem, which applies to a spherical shell, underestimates and poorly approximates the field of a ring.

Integrating the ring potential in Eq. (C.2) over the many rings that comprise a disk of areal mass density $\sigma(\rho)$ gives the disk potential in the plane of the disk as

$$\Phi(r) = -4G \int_0^\infty d\rho\, \sigma(\rho) K(k) / (1 + r/\rho) , \qquad (C.4)$$

where the modulus is $k(\rho) \equiv 2(r\rho)^{1/2} / (r + \rho)$, and the gravitational field in the plane of the disk is $g(r) = -d\Phi / dr$. (The singularity in $K(k)$ at $r = \rho$ is logarithmic, as $K(\rho) \sim \ln(4 / |1 - r/\rho|)$.) The potential of the thin disk has a

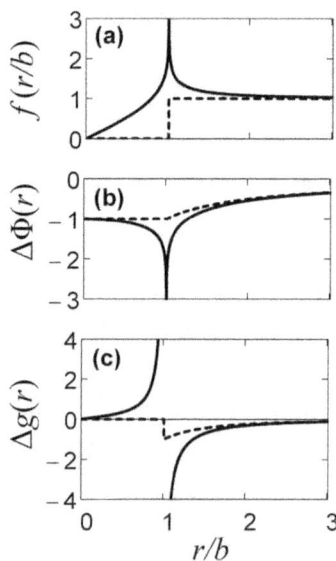

Fig. C.2. Field quantities at r for thin ring of mass Δm and radius b (solid curves) and thin spherical shell of mass Δm and radius b (dashed curves) vs. r/b: (a) Enhancement factor; (b) potential normalized to $G\Delta m / b$; and (c) field normalized to $G\Delta m / b^2$.

discontinuity in slope at the disk boundary. The gravitational field is discontinuous at the boundary. Figure C.3 compares the Newtonian potential and gravitational field of a thin disk and a sphere, each having uniform density and radius 1.

Consider a disk galaxy having a nearly exponential areal mass density profile,

$$\sigma(\rho) = \sigma_0 \exp(-\rho / h), \tag{C.5}$$

with some constant scale length h and areal mass density σ_0 at the center. With this density profile, the mass of a disk with radius 1 is

$$m_0 = 2\pi\sigma_0 h^2 \left[1 - (1 + 1/h)\exp(-1/h)\right]. \tag{C.6}$$

For a short scale length ($h \ll 1$) relative to the disk radius 1, $m_0 \approx 2\pi h^2 \sigma_0$. For a long scale length ($h \gg 1$), $m_0 \approx \pi\sigma_0$. Figure C.4 shows the Newtonian potential and gravitational field for thin disks with exponential density profiles, the same mass, and the same radius 1, but with different scale lengths.

With the Newtonian gravitational field $g(\rho)$ of a disk, the velocity of a rotation curve is defined as $v(\rho) = [-\rho g(\rho)]^{1/2}$. Figure C.5 shows the rotation curves calculated from Eqs. (C.4) and (C.5) for several truncated exponential disks of different scale heights. Van der Kruit and Searle found that exponential disks are often truncated at about four scale lengths [33], corresponding to $h = 0.25$ in the normalization of Fig. C.5. This figure shows that such exponential disks have flat rotation curves according to unmodified Newtonian dynamics. Such truncated exponential disks often show evidence of a cusp in the velocity rotation curves at the density truncation, a feature seen in the rotation curves in Fig. C.5 as well.

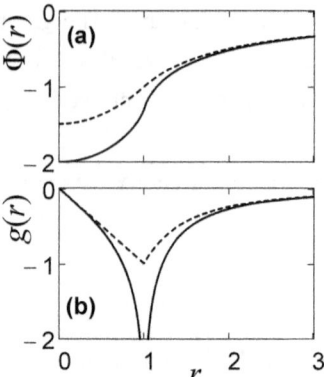

Fig. C.3. (a) Newtonian potential and (b) Newtonian gravitational field normalized to Gm_0 vs. range normalized to radius of disk (solid curve) and sphere (dashed curve), each having uniform density, mass m_0, and radius 1.

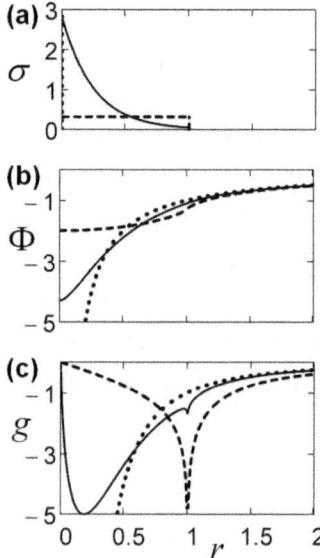

Fig. C.4. (a) Exponential areal mass density profiles, (b) Newtonian potentials, and (c) Newtonian gravitational fields normalized to $Gm_0 = 1$ vs. range normalized to radius of disks, each having the same mass and same radius 1, but with scale heights $h = 0$ (dotted), $h = 0.25$ (solid), $h = \infty$ (dashed).

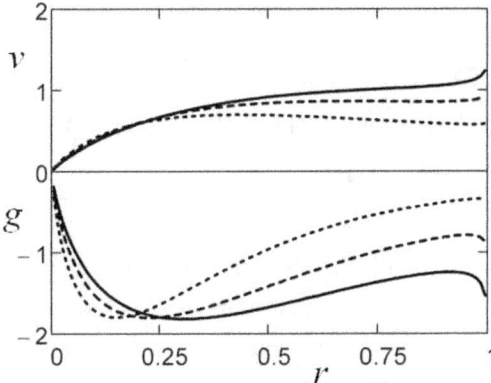

Fig. C.5. Normalized velocity rotation curves (upper) and Newtonian gravitational fields (lower) normalized to $Gm_0 = 1$ *vs*. range normalized to radius of disks, each having the same central areal mass density $\sigma_0 = 1$ and same radius 1, but with scale heights $h = 0.4$ (solid), $h = 0.3$ (dashed), $h = 0.2$ (dotted).

Appendix D
Gravitational Field of a Particle in Arbitrary Relativistic Motion

This appendix presents the derivation from [4] of the gravitational field of a particle undergoing arbitrary relativistic motion. The expression for gravitational field is exact to all orders of source velocity, but is valid only for a weak gravitational field of the source, that is, for a gravitational potential of the source much less than its specific kinetic energy, so that the source is a gravitationally unbound quadrupole.

Kopeikin [59] recognized that solving for the *dynamic* field of a relativistic particle in arbitrary motion requires a Liénard-Wiechert "retarded solution" approach. By using the retarded Green's function to solve the linearized field equations in the weak-field approximation, [59] calculated the exact "retarded Liénard-Wiechert tensor potential" of a relativistic particle of rest mass m as

$$h_{\mu\nu}(\mathbf{r},t) = \frac{-4Gm}{c^4} \int \frac{S_{\mu\nu}(t')}{\gamma(t')R(t')} \delta\left(t' + \frac{R(t')}{c} - t\right) dt' = \frac{-4Gm}{c^4} \left\{\frac{S_{\mu\nu}}{\gamma\kappa R}\right\}_{ret}. \quad (D.1)$$

In the weak-field approximation used in Eq. (D.1), the metric tensor was linearized as $g_{\mu\nu} = \eta_{\mu\nu} + h_{\mu\nu}$, where $\eta_{\mu\nu}$ is the Lorentz metric. With respect to Fig. 11.3, $S_{\mu\nu} \equiv u_\mu u_\nu - c^2 \eta_{\mu\nu}/2$ is a source tensor, with pressure and internal energy neglected; $u_\mu = \gamma[c, \mathbf{u}]$ is the 4-velocity of the source mass m; $\mathbf{u}$ is the 3-velocity; $\boldsymbol{\beta} = \mathbf{u}/c$ is the normalized 3-velocity; $\gamma = (1-\beta^2)^{-1/2}$ is the Lorentz (relativistic) factor; $\mathbf{R} = \mathbf{r} - \mathbf{s}'$ is the displacement vector from the source position $\mathbf{s}'(t')$ to the test particle at $\mathbf{r}(t)$; $\mathbf{n} = \mathbf{R}/R$ is a unit vector; the delta function provides the retarded behavior required by causality; the factor $\kappa \equiv 1 - \mathbf{n} \cdot \mathbf{u}/c$ is the derivative with respect to t' of the argument of the delta function, $t' + [R(t')/c] - t$; and the quantity in brackets $\{\ \}_{ret}$ and all primed quantities are to be evaluated at the retarded time $t' = t - R'/c$.

Equation (D.1) is the starting point in this appendix for the exact calculation of the gravitational field of a relativistic particle from the tensor potential. In the weak-field approximation, the retarded tensor potential of Eq. (D.1) is exact, even for relativistic velocities of the source. And since the tensor potential is linear, the field to be derived from it in this appendix is easily generalized to ensembles of particles and to continuous source distributions.

To derive an exact expression for the gravitational field from Eq. (D.1) by the Liénard-Wiechert formalism [60], a 'scalar potential' is defined as

$$\Phi(\mathbf{r},t) \equiv \frac{c^2 h_{00}}{2} = -Gm \int \frac{\alpha'}{R'} \delta\left(t' + \frac{R'}{c} - t\right) dt' \ , \tag{D.2}$$

and the components of a 'vector potential' are defined as

$$A^i(\mathbf{r},t) \equiv c^2 h_0^{\ i} = -4Gm \int \frac{(\beta^i)' \gamma'}{R'} \delta\left(t' + \frac{R'}{c} - t\right) dt' \ , \tag{D.3}$$

where $\alpha \equiv 2\gamma - 1/\gamma$, and $i = 1,2,3$. Then from the geodesic equation, the equation of motion of a test particle *instantaneously at rest* at $(\mathbf{r},t)$ in a weak field is

$$\frac{d^2\mathbf{r}}{dt^2} = -\nabla\Phi(\mathbf{r},t) - \frac{1}{c}\frac{\partial \mathbf{A}(\mathbf{r},t)}{\partial t} \ . \tag{D.4}$$

Since the gradient operation in Eq. (D.4) is equivalent to $\nabla \to \mathbf{n}\partial / \partial R$, the contribution of the 'scalar potential' to the gravitational field can be written as

$$-\nabla\Phi = -Gm \int \left[\frac{\alpha\mathbf{n}}{R^2}\delta\left(t' + \frac{R'}{c} - t\right) - \frac{\alpha\mathbf{n}}{cR}\tilde{\delta}\left(t' + \frac{R'}{c} - t\right) \right] dt' \ . \tag{D.5}$$

The contribution of the 'vector potential' is

$$-\frac{1}{c}\frac{\partial \mathbf{A}}{\partial t} = \frac{-4Gm}{c} \int \frac{\gamma\boldsymbol{\beta}}{R}\tilde{\delta}\left(t' + \frac{R'}{c} - t\right) dt' \ , \tag{D.6}$$

where $\tilde{\delta}$ is the derivative of the delta function with respect to its argument. If the variable of integration is changed to $f(t') = t' + R'/c$, then integrating by parts on the derivative of the delta function, and combining Eqs. (D.4) to (D.6) gives

$$\mathbf{g}(\mathbf{r},t) = -Gm\left\{ \frac{\alpha\mathbf{n}}{\kappa R^2} + \frac{1}{c\kappa}\frac{d}{dt'}\left(\frac{\alpha\mathbf{n} - 4\gamma\boldsymbol{\beta}}{\kappa R} \right) \right\}_{\text{ret}} \ . \tag{D.7}$$

To calculate the time derivatives in Eq. (D.7), the following relations from [60] are used,

$$\frac{1}{c}\frac{d\mathbf{n}}{dt'} = \frac{\mathbf{n}\times(\mathbf{n}\times\boldsymbol{\beta})}{R} = \frac{(\mathbf{n}\cdot\boldsymbol{\beta})\mathbf{n}-\boldsymbol{\beta}}{R} \quad , \tag{D.8}$$

$$\frac{1}{c}\frac{d(\kappa R)}{dt'} = \beta^2 - \mathbf{n}\cdot\boldsymbol{\beta} - \frac{R(\mathbf{n}\cdot\dot{\boldsymbol{\beta}})}{c} \quad , \tag{D.9}$$

where an overdot denotes differentiation with respect to t'. Applying these relations, Eqs. (D.8) and (D.9), to Eq. (D.7) gives the relativistically exact (weak) retarded gravitational field of a moving source on a test particle instantaneously at rest at $(\mathbf{r},t)$ as Eq. (11.2),

$$\mathbf{g}(\mathbf{r},t) = -Gm\left\{\frac{(\alpha/\gamma^2)\mathbf{n}+[(2\gamma+1/\gamma)\kappa-4/\gamma]\boldsymbol{\beta}}{\kappa^3 R^2}\right.$$
$$\left.+\frac{(\mathbf{n}\cdot\dot{\boldsymbol{\beta}})(\alpha\mathbf{n}-4\gamma\boldsymbol{\beta})+\kappa(\dot{\alpha}\mathbf{n}-4\dot{\gamma}\boldsymbol{\beta}-4\gamma\dot{\boldsymbol{\beta}})}{c\kappa^3 R}\right\}_{\text{ret}} \tag{D.10}$$

The radial component of the gravitational 'velocity field' can change sign at sufficiently high source velocity and repel masses within a narrow cone. This behavior gives rise to a gravitational repulsion at relativistic velocities, first discovered for axial motion in strong fields by [61, 62], and generalized by [4, 63] for arbitrary relativistic motion of the source in the weak-field approximation.

Equation (D.10) is the gravitational (or inertial) field that would be observed by a distant inertial observer to act on a test particle at rest at $(\mathbf{r},t)$. If the test particle moves with velocity $\mathbf{v}$, then the inertial field measured by the moving test particle has additional 'gravimagnetic' terms. The calculations of inertial radiation in Ch. 11 apply in the rest frame of the test particle and use the impulse approximation, which has the test particle remaining at rest during the time the inertial field of the moving particle acts upon it. (Any difference in fields caused by motion induced in the test particle by the source is of the same order as terms that have already been neglected in the weak-field approximation.)

Appendix E
General Statistical Properties of a Noise Field

This appendix calculates the probability distribution of the amplitude of a noise field within a specified frequency band. The calculation begins by considering a large number N of field modes, each having the same frequency and unit amplitude, but random phase. Then the calculation considers the effects of a distribution of mode amplitudes and of a distribution of mode frequencies.

Consider a plane-wave field,

$$f(z,t) = \sum_{i=1}^{N} f_i(z,t) = \sum_{i=1}^{N} a_i \cos\left[\omega_i(t - z/c) + \phi_i\right],$$ (E.1)

propagating with velocity c in the z direction and comprising N field modes, where a_i is the amplitude, ω_i is the angular frequency, and ϕ_i is the phase of the ith field mode.

To begin, let each field mode have equal frequency and unit amplitude, but random phase. Then a measurement of the field at any arbitrary spacetime point yields a field strength,

$$f = \sum_{i=1}^{N} \cos(\phi_i),$$ (E.2)

where the ϕ_i are random uniformly from 0 to 2π. A large number M of such field-strength measurements will be uncorrelated if the times or locations of the measurements are unrelated.

As shown in Fig. E.1(a), uncorrelated field-strength measurements will be normally distributed about zero field strength. That is, the probability that a measurement of field strength will have a magnitude between $|f|$ and $|f| + d|f|$ is

$$P(|f|)d|f| = \frac{2}{\sqrt{\pi N}} \exp\left(-\frac{f^2}{N}\right) d|f|.$$ (E.3)

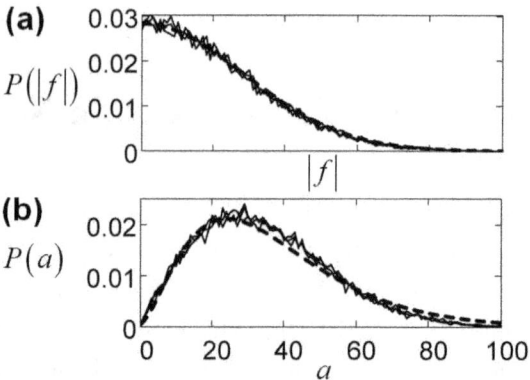

Fig. E.1. Probability for 1600 measurements each, with uniform, linear, and negative-binomial distributions of field amplitudes (3 solid curves), that a measurement of: (a) field strength f will have amplitude between $|f|$ and $|f| + d|f|$; (b) field amplitude a will have magnitude between a and $a + da$. Dashed curve in (a) is normal probability distribution, and in (b) is best-fit negative binomial distribution.

For this appendix, the probability distribution of field amplitudes a is of greater interest than the distribution of field strength measurements. The field amplitude is found from a particular measurement of field strength by

$$a = \left[\left(\sum_{i=1}^{N} \cos(\phi_i) \right)^2 + \left(\sum_{i=1}^{N} \sin(\phi_i) \right)^2 \right]^{1/2}. \tag{E.4}$$

The sum over cosines in Eq. (E.4) is correlated with the sum over sines. Equation (E.4) may be written as

$$a = \left[N + 2 \sum_{j>i}^{N} \sum_{i=1}^{N} \cos(\phi_i - \phi_j) \right]^{1/2}. \tag{E.5}$$

The double sum over cosines in Eq. (E.5) must be at least weakly correlated, even for large N, because the double sum has $N(N-1)/2$ terms, each of which can be negative, yet Eq. (E.4) shows that the quantity inside the square root must be positive. In the limit $N \to \infty$, however, the terms in the double sum become uncorrelated, and the peak field power density as $N \to \infty$,

$$a^2 \rightarrow N + \sum_{k=1}^{N(N-1)} \cos(\phi_k), \tag{E.6}$$

has a probability distribution as $N \rightarrow \infty$,

$$P_\infty(a^2) \rightarrow (\pi^{1/2} N)^{-1} \exp\left[-(a^2 - N)^2 / N^2\right]. \tag{E.7}$$

That is, the peak power density a^2 of the noise field has a mean value N, and a standard deviation $N / 2^{1/2}$ in the limit $N \rightarrow \infty$.

For a large number of modes N, but not so large that N effectively approaches infinity, a simple fit for the probability distribution of the field amplitude a is the negative-binomial distribution [64],

$$P(a) \approx \frac{\Gamma(a+r)}{\Gamma(a+1)\Gamma(r)} p^r (1-p)^a, \tag{E.8}$$

where Γ is the gamma function, r is the shape parameter, and $p = r / N^{1/2}$ is the spread parameter. The mean of the negative-binomial distribution is

$$\bar{a} = r(1-p)/p, \tag{E.9}$$

or $\bar{a} \rightarrow N^{1/2}$ in the limit $N \rightarrow \infty$. The standard deviation is

$$\sigma(a) = [r(1-p)]^{1/2} / p, \tag{E.10}$$

or $\sigma(a) \rightarrow (N/r)^{1/2}$ in the limit $N \rightarrow \infty$. The shape parameter r that gives a best fit of the negative-binomial distribution to the histogram of $P(a)$ is a very slowly varying function of N. For $N = 1600$, the best fit of $P(a)$ to the histograms of measurements in Fig. E.1(b), is obtained for $r \approx 3$. For $N = 1600$ modes, with uniform mode amplitude $a_i = 1$, the mean field amplitude is $\bar{a} = 37$, and the standard deviation is $\sigma(a) = 22$.

Next, suppose that the modes, instead of all having unit amplitude, have a distribution of amplitudes. A combination of modes having a negative-binomial distribution of mode amplitudes results in a combined field amplitude with the same negative-binomial probability distribution. But so does just about any reasonable distribution of mode amplitudes, including those shown in Fig. E.2.

148

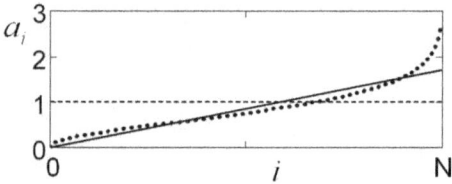

Fig. E.2. Distributions of mode amplitudes a_i vs. mode number i: negative binomial (dotted curve); uniform (dashed line); and linear (solid line).

To demonstrate, mode amplitudes a_i are assigned a negative-binomial distribution by means of the following algorithm. The set of mode amplitudes is partitioned into bins, where the nth bin contains all modes with amplitudes a_i, such that $n-1 < a_i \leq n$. From Eq. (E.8), the number of modes in the nth bin is $NP(n)$, rounded to the nearest integer. Each of these modes in the nth bin is assigned an amplitude $a_i = n\bar{a} / N$. This negative-binomial distribution of mode amplitudes is shown as the dotted curve in Fig. E.2.

As seen in Figs. E.1(a) and E.1(b), when the mode amplitudes are assigned a negative-binomial distribution in this way, the probability distributions of field measurements, $P(|f|)$, and of field amplitudes, $P(a)$, are statistically indistinguishable from the probability distributions for uniformly distributed mode amplitudes.

To demonstrate that the negative-binomial distribution is not unique in this regard, a linear distribution of mode amplitudes, shown in Fig. E.2, is also tested. As seen in Figs. E.1(a) and E.1(b), the linear distribution of mode amplitudes (with the same mean of a_i as the negative-binomial distribution) results in the same statistically indistinguishable probability distributions, $P(|f|)$ and $P(a)$, as do the uniform and negative-binomial distributions of mode amplitudes.

In summary, when a large number of modes N, each having equal frequency, unit amplitude, and random phase, are added together, the resultant field has a mean amplitude $N^{1/2}$ and a wide standard deviation comparable to the mean. If the modes have any distribution of amplitudes statistically equivalent to a uniform distribution, as for example the distributions shown in Fig. E.2, then the resultant field is statistically indistinguishable and has the same mean and standard deviation as the field comprising a uniform distribution of mode amplitudes.

Next, this appendix calculates the effects of a distribution of mode frequencies. A continuous measurement of the plane-wave field of Eq. (E.1) at a fixed point in space yields a field strength,

$$f(t) = \sum_{i=1}^{N} a_i \cos(\omega_i t + \phi_i),$$ (E.11)

where the ϕ_i are random uniformly from 0 to 2π, as before, but both the mode amplitudes a_i and the mode frequencies ω_i may have some distribution of values. Here, a narrowband distribution of mode frequencies is of interest, not because the noise spectrum is narrowband, but because a resonant detector used as a diagnostic to probe the spectrum may be narrowband. And a narrowband (high-Q) detector has a significant response only to those mode frequencies that lie within its bandwidth.

First, consider a linear distribution of mode frequencies, ω_1 through ω_N, having a fractional bandwidth Δ about a central frequency ω_0,

$$\omega_i = \omega_0 + \left(\frac{i-1}{N-1} - \frac{1}{2} \right) \Delta \omega_0.$$ (E.12)

For a uniform distribution of mode amplitudes, $a_i = 1$, and a fractional bandwidth $\Delta = 0.1$, typical field amplitudes vs. time are shown by circles in Figs. E.3(a) and E.3(b) for $N = 2$ and $N = 1600$, respectively.

For $N = 2$, the field is the beat wave, $2\cos(\omega_0 t)\cos[(\Delta/2)\omega_0 t]$. The beat wave is a product of a sinusoidal function $\cos(\omega_0 t)$ with wave period $2\pi/\omega_0$ and an envelope function with modulation period $2\pi/\Delta\omega_0$, so that the duration of one beat-wave modulation is $1/\Delta$ wave periods, as seen in Fig. E.3(a).

For $N = 1600$, measurements of field amplitude vs. time are expected to have the same distribution over many measurements and over long times as do many uncorrelated amplitude measurements of a linear combination of 1600 modes, namely, the negative-binomial distribution of Eq. (E.8). In fact, the mean field amplitude in Fig. E.3(b) over long times is about the same as that given by Eq. (E.9) for $N = 1600$, namely $\bar{a} = 37$. Over long times, many measurements of amplitude vs. time at a fixed location are statistically equivalent to many uncorrelated measurements of amplitude.

For the same reason, it does not matter if the mode frequencies are distri-

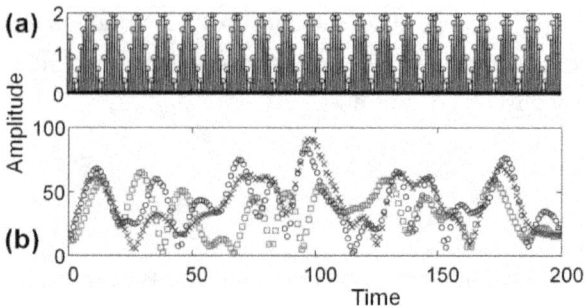

Fig. E.3. Symbols indicate peak amplitude in each wave period *vs.* time in wave periods for modes with fractional bandwidth 0.1: (a) $N = 2$ modes; (b) $N = 1600$ modes. Circles indicate linear frequency distribution within bandwidth with unit mode amplitude. Crosses indicate linear frequency distribution with negative-binomial distribution of mode amplitudes. Squares indicate random frequency distribution with unit mode amplitude.

buted linearly or are random uniformly from $\omega_0 - \Delta\omega_0 / 2$ to $\omega_0 + \Delta\omega_0 / 2$.

Nor does it matter if the modes all have unit amplitude or have the negative-binomial distribution of amplitudes given by Eq. (E.8). As shown in Fig. E.3(b), in any case the field amplitudes *vs.* time are statistically indistinguishable and the mean field amplitudes and standard deviations measured over long times are the same.

Comparing Fig. E.3(a) with E.3(b), one sees that the modulation of a beat wave comprising two modes is $1/\Delta$ wave periods long, but the 'modulation' of a field comprising many modes approaches twice the duration, or $2/\Delta$ wave periods long, depending on how one counts 'modulations'. This effective doubling of the modulation period can perhaps be understood as follows. A linear combination of N field modes of unit amplitude, with frequencies distributed linearly over a bandwidth $\Delta\omega_0$, is equivalent to a linear combination of $N/2$ beat waves, each comprising a pair of modes with a difference frequency $\Delta\omega_0 / 2$. That is, the difference frequency of the $N/2$ beat waves is half the bandwidth, which means the modulation period of each of the $N/2$ beat waves is twice as long.

In summary, when a large number of modes N, each having unit amplitude and random phase, but with frequencies distributed linearly or randomly over a fractional bandwidth Δ , are added together, the resultant field measured continuously at any fixed location has a mean amplitude $N^{1/2}$ and a wide standard deviation comparable to the mean. The depth and duration of the field modulations are irregular, but the mean duration of the modulations is less than or about $2/\Delta$ wave periods long. The fields are statistically indistinguishable whether the modes have unit amplitude or the statistically equivalent amplitude distributions shown in Fig. E.2.

Appendix F
Conjectural Cosmogony

An objection to the concept first proposed by Einstein of a static, homogeneous universe is that no physical mechanism is offered to explain the origin of such a universe. Absent such a physical mechanism, conceptual problems arise that mainly relate to the apparent age of the universe. According to [65], for example, the apparent age of our universe, if static, does not agree with the limited lifetimes of stars and stellar systems or with the degree of large-scale homogeneity. Such large-scale homogeneity is gravitationally unstable and could not persist much longer than the lifetime of a solar-mass star in a non-expanding universe [65]. Such a cosmology, the static, homogeneous universe proposed by Einstein, appears unaccountably through this reasoning to have been created relatively recently, everywhere, all at once.

This appendix discusses whether cosmogony, the inquiry into the origination of our universe, can ever advance beyond the conjectural stage through the scientific method.

Figure F.1 depicts several examples of cosmological models that have been proposed. Some models, like the Egyptian Old Kingdom circular mound of creation, *ca.* 2780 – 2250 B.C.E., in Fig. F.1(a) and the Big Bang in Fig. F.1(b), purport to be etiological and presume an origin. Others, like the Einstein static, homogeneous universe represented in Fig. F.1(c), do not.

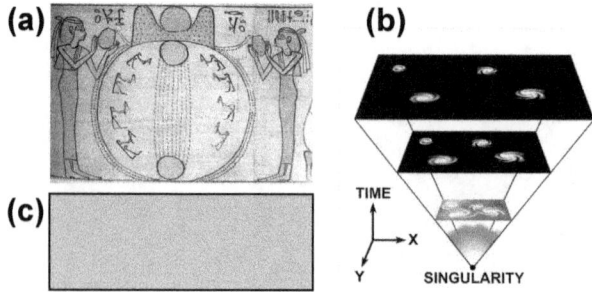

Fig. F.1. Examples of cosmologies that presume an origin: (a) Egyptian circular mound of creation and (b) Big Bang; and those that do not: (c) Einstein static universe.

Proponents of the *principle of sufficient reason*, usually attributed to Leibniz, that everything must have a cause, believe that there should be a physical mechanism that explains the origin of the universe, and they favor cosmological models that presume an origin. *Observationalists*, on the other hand, question whether the origin of the universe can be regarded as a subject for scientific inquiry at all, citing the absence of means of empirical falsification in etiological models.

For a model to be falsifiable means that an experiment can be performed, one possible outcome of which is that the result conflicts with the underlying hypothesis and the model must be rejected. Karl Popper is generally credited with replacing the concept of verification in the scientific method with the concept of *falsification*, the formal attempt to disprove hypotheses rather than prove them [5]. Falsification is introduced into the scientific method as a way to reduce *confirmation bias*, the observer's bias towards confirming his expectations and views of the world, that is, his bias to see what he <u>expects</u> to see. Confirmation bias leads to the tendency to favor information that confirms one's beliefs. Confirmation bias also leads to other kinds of bias, like *processing bias*, which is the tendency to process and present data in such a way as to see what one <u>wants</u> to see, blurring the lines between 'observing' and 'drawing conclusions'.

Figure F.2 illustrates schematically a composite version of the scientific method. The process begins when a question arises and a potential answer becomes a conjecture. By Popper's standards, a hypothesis can be formulated from the conjecture only if the conjecture is consistent with all available observations <u>and</u> is falsifiable. That is, at least one test can be proposed with a potential outcome being that the hypothesis is discarded. After being subjected to many tests, all of which are capable of falsifying the hypothesis, but none of which does falsify the hypothesis, the hypothesis may come to be regarded as a theory.

A recent example of a hypothesis successfully formulated from a conjecture is the existence of the Higgs boson. The conjecture of a new particle in the predicted mass range was not inconsistent with any available observations. Once the conjecture became falsifiable by tests at the Large Hadron Collider, it qualified as a hypothesis. An example of a theory is Einstein's theory of relativity. Over the better part of a century, it has been subjected to countless tests, any of which could have falsified the theory.

154

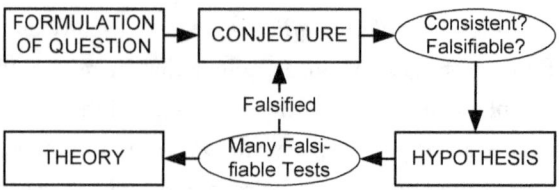

Fig. F.2. Scientific method rigorously screens conjectures and unceasingly attempts to falsify hypotheses.

These examples stand in contrast to models of the origin of the universe. What falsifiable tests have ever been proposed, for example, that could result in the Big Bang model of origination being universally discredited and discarded? Instead we see the *persistence of discredited beliefs*, the phenomenon, usually explained by confirmation bias, of some beliefs remaining after initial evidence is removed. Arguably, any of a number of observations, especially when considered together, could have sufficed to falsify a Big Bang cosmogony, such as:

• *Discovery of isotropy.* In no direction does there appear an edge or a preferred center of the universe that might have been the site of a Big Bang, and consequently "no one has thought of a way to adduce objective physical evidence that such an event really happened [65]."

• *Discovery of 'acceleration'.* The universe cannot be observed to have a spherically symmetric acceleration of its expansion, except about a preferred center, in violation of the Copernican principle. See Fig. 5.2.

• *Discovery of homogeneity.* Homogeneous, constant density profiles only ever appear under conditions of static equilibrium or unaccelerated expansion and never in dynamic strong explosions. See, *e.g.*, [66].

• *Inconsistent energy density.* The <u>observable</u> mass and energy density of the universe is inconsistent by more than an order of magnitude with the Big Bang model.

The effectiveness of supplanting verification with falsification in the scientific method may be measured in part by attempts, motivated by confirmation bias, to replace falsification with verification. For example, the searches for dark energy and dark matter have thus far been attempts at verification, not falsification.

While cosmogony, from the Egyptian circular mound of creation to more recent models such as inflation, may not yet have advanced beyond conjecture, at least by the standards specified here, the same does not apply of course to cosmology, or its ancient progenitor, astronomy. The science of astronomy has been esteemed for millennia. For example, the Talmud quotes Shimon Bar Kappara, *ca.* 180 – 220 CE, as saying [67], "Anyone who knows how to calculate seasons and constellations and he does not calculate, Scripture states about him: And the work of God they do not regard and the action of His hands they do not see."

References

1. Einstein, A.: Letter to E. Mach, Zurich, 25 June 1913; reproduced and translated in [2].

2. Misner, C.W., Thorne, K.S., Wheeler, J.A.: Gravitation. Freeman & Co., NY (1973), pp. 544–545.

3. Ciufolini, I., Wheeler, J.A.: Gravitation and Inertia. Princeton U. Press, Princeton, NJ, (1995), p. 299.

4. Felber, F. S.: Weak 'antigravity' fields in general relativity. arXiv:gr-qc/0505098v3. http://arxiv.org/ftp/gr-qc/books/0505/0505098.pdf (2005). Accessed Oct. 2015.

5. Popper, K.R.: Logik der Forschung (in German). Springer Verlag, Vienna (1935); The logic of scientific discovery. Routledge, NY (2002).

6. NASA, prepared by STSci. http://hubblesite.org.

7. Polyanin, A.D., Zaitsev, V. F.: Handbook of exact solutions for ordinary differential equations, 2nd Ed. Chapman & Hall, CRC Press, NY (2003).

8. Einstein, A., Infeld, L.: The evolution of physics, from early concepts to relativity and quanta. Simon & Schuster, NY (1966), pp. 242–243.

9. Felber, F. S.: Dipole gravity waves from unbound quadrupoles. arXiv:1002.0351v2 [physics.gen-ph]. http://arxiv.org/ftp/arxiv/books/1002/1002.0351.pdf (2010).

10. Weinberg, S.: Cosmology. Oxford U. Press, Oxford (2008), pp. 28–31.

11. Freedman, W. L., et al.: Final results from the Hubble Space Telescope Key Project to measure the Hubble constant. Astrophys. J. 553, 47 (2001).

12. Perlmutter, S., et al.: Discovery of a supernova explosion at half the age of the universe. Nature 391, 51–54 (1998).

13. Perlmutter, S., et al.: Measurements of Ω and Λ from 42 high-redshift supernovae. Astrophys. J. 517, 565 (1999).

14. Riess, A. G., et al.: Observational evidence from supernovae for an accelerating universe and a cosmological constant. Astron. J. 116, 1009–1038 (1998).

15. Schmidt, B., et al.: The high-Z supernova search: Measuring cosmic decel-

eration and global curvature of the universe using Type Ia supernovae. Astrophys. J. 507, 46 (1998).

16. Hicken, M., *et al.*: Improved dark energy constraints from ~100 new CfA supernova Type Ia light curves. Astrophys. J. 700, 1097 (2009).

17. Conley, A., *et al.*: Is there evidence for a Hubble bubble? The Nature of Type Ia supernova colors and dust in external galaxies. Astrophys. J. 664, L13–L16 (2007).

18. Moss, A., Zibin, J., Scott, D.: Precision cosmology defeats void models for acceleration. Physical Review D 83, 103515 (2011).

19. Nielsen, J.T., Guffanti, A., Sarkar, S.: Marginal evidence for cosmic acceleration from Type Ia supernovae. arXiv:1506.01354v2 [astro-ph.CO]. http://arxiv.org/pdf/1506.01354v2.pdf. Accessed Oct. 2015.

20. Yu, H-R, Zhang, T-J, Pen, U-L: Method for Direct Measurement of Cosmic Acceleration by 21-cm Absorption Systems. Phys. Rev. Lett. 113, 041303 (2014).

21. Sandage, A.: The change of redshift and apparent luminosity of galaxies due to the deceleration of selected expanding universes. Astrophys. J. 136, 319 (1962).

22. Loeb, A.: Direct measurement of cosmological parameters from the cosmic deceleration of extragalactic objects. Astrophys. J. 499, L111–L114 (1998).

23. Anderson, J. D., *et al.*: Study of the anomalous acceleration of Pioneer 10 and 11. Phys. Rev. D 65, 082004 (2002).

24. Nieto, M.M., Anderson, J.D.: Search for a solution of the Pioneer anomaly. Contemp. Phys. 48, 41 (2007).

25. Iorio, L.: Can the Pioneer anomaly be of gravitational origin? A phenomenological answer. Found. Phys. D 37, 897–918 (2007).

26. Iorio, L.: Orbital effects of a time-dependent Pioneer-like anomalous acceleration. Mod. Phys. Lett. A 27, 1250071 (2012).

27. Bennett, C.L., *et al.*: First-year Wilkinson Microwave Anisotropy Probe (WMAP) observations: Preliminary maps and basic results. Astrophys. J. Suppl. 148, 1 (2003).

28. Smoot, G.F., *et al.*: Preliminary results from the COBE differential microwave radiometers: Large angular scale isotropy of the cosmic microwave background. Astrophys. J. Part 2 – Letters 371, L1–L5 (1991).

29. Kogut, A., *et al.*: Dipole anisotropy in the COBE differential microwave

radiometers first-year sky maps. Astrophys. J. 419, 1 (1993).

30. Turyshev, S.G., et al.: Support for temporally varying behavior of the Pioneer anomaly from the extended Pioneer 10 and 11 Doppler data sets. Phys. Rev. Lett. 107, 081103 (2011).

31. Turyshev, S.G., et al.: Support for the thermal origin of the Pioneer anomaly. Phys. Rev. Lett. 108, 241101 (2012).

32. Sancisi, R., van Albada, T.S.: In: Kormendy, J., Knapp, G.R. (eds.) IAU Symp. 117: Dark matter in the universe, p. 67. Reidel, Dordrecht (1987)

33. Sanders, R.H.: The dark matter problem, a historical perspective. University Press, Cambridge, U.K. (2010).

34. Tully, R.B., Fisher, J.R.: A new method of determining distances to galaxies. Astron. Astrophys. 54, 661–673 (1977).

35. Milgrom, M.: A modification of Newtonian dynamics as a possible alternative to the hidden matter hypothesis. Astrophys. J. 270, 365–370 (1983).

36. Famaey, B., McGaugh, S.S.: Modified Newtonian dynamics (MOND): Observational phenomenology and relativistic extensions. Living Rev. Relativ. 15, 10 (2012).

37. Milgrom, M.: The MOND paradigm of modified dynamics. Scholarpedia 9, 31410 (2014).

38. Milgrom, M.: MOND theory. Canad. J. Phys. 93, 107 (2015).

39. Sanders, R.H., McGaugh, S.S.: Modified Newtonian dynamics as an alternative to dark matter. Ann. Rev. Astron. Astrophys. 40, 263–317 (2002).

40. Freeman, K.C.: On the disks of spiral and S0 Galaxies. Astrophys. J. 160, 811–830 (1970).

41. Faber, S.M., Jackson, R.E.: Velocity dispersions and mass-to-light ratios for elliptical galaxies. Astrophys. J. 204, 668–683 (1976).

42. Emden, R.: Gaskugeln. Teubner, Leipzig, Berlin (1907).

43. Kalnajs, A.J.: Internal Kinematics and Dynamics of Galaxies. In Athanassoula, E. (ed.) IAU Symp. 100, p. 87. Reidel, Dordrecht (1983).

44. Allen, R.J., Shu, F.H.: The extrapolated central surface brightness of galaxies. Astrophys. J. 227, 67–72 (1979).

45. Clowe, D., et al.: A direct empirical proof of the existence of dark matter. Astrophys. J. 648, L109–L113 (2006).

46. Clowe, D., Randall, S.W., Markevitch, M.: Catching a bullet: Direct evidence for the existence of dark matter. Nucl. Phys. B - Proc. Suppl. 173,

28–31 (2007).

47. Markevitch, M., *et al.*: A textbook example of a bow shock in the merging galaxy cluster 1E 0657–56. Astrophys. J. 567, L27–L31 (2002).

48. Milgrom, M.: Ultra-diffuse cluster galaxies as key to the MOND cluster conundrum. Mon. Not. R. Astron. Soc. 454, 3810 (2015).

49. Angus, G.W., McGaugh, S.S.: The collision velocity of the bullet cluster in conventional and modified dynamics. Mon. Not. R. Astron. Soc. 383, 417–423 (2008).

50. Felber, F. S.: Source-region calculation of dipole power. arXiv:physics/0510115v1   [physics.plasm-ph]. http://arxiv.org/ftp/physics/books/0510/0510115.pdf (2005).

51. Felber, F. S.: Dipole radio-frequency power from laser plasmas with no dipole moment. Appl. Phys. Lett. 86, 231501 (2005).

52. Felber, F. S.: Exact 'antigravity-field' solutions of Einstein's equation. arXiv:0803.2864v4   [physics.gen-ph]. http://arxiv.org/ftp/arxiv/books/0803/0803.2864.pdf (2008).

53. Felber, F. S.: New exact dynamic field solutions of Einstein's equation. Bull. Am. Phys. Soc. 54, 241 (2009).

54. Ohanian, H., Ruffini, R.: Gravitation and Spacetime, 2nd Ed. W. W. Norton & Co., NY (1994).

55. Nelson, E.: Derivation of the Schrödinger equation from Newtonian mechanics. Phys. Rev. 150, 1079 (1966).

56. Einstein, A.: Letter to M. Born, 4 Dec. 1926. In Born, M. (ed.) The Born-Einstein letters, p. 91. Walker, NY (1971).

57. Sakharov, A.D.: Vacuum quantum fluctuations in curved space and the theory of gravitation. Sov. Phys. - Doklady 12, 1040 (1968).

58. De, S., Chakrabarty, S.: Schrödinger equation of a particle in an uniformly accelerated frame and the possibility of a new kind of quanta. Mod. Phys. Lett. A 30, 1550182 (2015).

59. Kopeikin, S.M., Schäfer, G.: Lorentz covariant theory of light propagation in gravitational fields of arbitrary-moving bodies. Phys. Rev. D 60, 124002 (1999).

60. Jackson, J.D.: Classical Electrodynamics. John Wiley & Sons, NY (1962); Classical Electrodynamics, 2nd Ed. John Wiley & Sons, NY (1975).

61. Hilbert, D.: Die Grundlagen der Physik, Zweite Mitteilung. In: Nachrichten

von der Königlichen Gesellschaft der Wissenschaften zu Göttingen, Mathematisch-Physikalische Klasse, pp. 53-76. University of Göttingen, Göttingen (1917).

62. Hilbert, D.: Grundlagen der Physik. Mathematische Annalen 92, 1–32 (1924).

63. Felber, F.S.: Exact relativistic 'antigravity' propulsion. AIP Conf. Proc. 813, 1374 (2006).

64. Weisstein, E.W.: Negative Binomial Distribution. MathWorld – A Wolfram Web Resource. http://mathworld.wolfram.com/NegativeBinomialDistribution.html. Accessed Oct. 2015.

65. Peebles, P.J.E.: Principles of Physical Cosmology. Princeton University Press, Princeton, NJ (1993).

66. Zel'dovich, Ya. B., Raizer, Yu. P.: Physics of shock waves and high-temperature hydrodynamic phenomena, Vol. II. Academic Press, NY (1967).

67. Talmud Bavli, Tractate Shabbos, Schottenstein Ed. Vol. 4. Mesorah Publications, NY (2000), p. 75a.

Index